D1598880

SERVANTS OF THE SWORD

DOUGLAS CLARK BAXTER

SERVANTS
OF THE SWORD
French Intendants of the Army
1630-70

UNIVERSITY OF ILLINOIS PRESS
Urbana Chicago London

Publication of this work was supported in part
by a grant from the Andrew W. Mellon Foundation

LIBRARY OF CONGRESS CATALOGING IN PUBLICATION DATA

Baxter, Douglas Clark, 1942–
    Servants of the sword.

    Bibliography: p.
    Includes index.
    1. France.   Armée—Management—History.   2. Inten-
dants.   3. Military   administration.   I. Title.
LLB71.B39   355.6'0944   75-40115
ISBN 0-252-00291-1

Contents

Preface

The historian studying seventeenth-century French history faces a difficult task. The amount of published material in this field is staggering, but much of it fails to provide answers to the questions that present-day historians find important. The problem of the army intendant is a good example. Despite the recent interest in administrative and institutional history, little study has been done on this subject. The most satisfactory attempt to treat the office occurs in the works on the Le Tellier administration by historian Louis André, although André was more concerned about Le Tellier's reforms than about the evolution of the intendancy. Other than this, there has been no real attempt to study this important office in the context of political history, specifically the expansion of royal authority in the military realm.[1]

There have been a few specialized studies of the missions of several seventeenth- and eighteenth-century intendants, but while interesting, they have reached few general conclusions about the nature of the office. For example, there are studies of the army intendancies of two subsequent secretaries of state for war: Sublet de Noyers and Michel Le Tellier. The only other examination of a seventeenth-century intendancy is that of the army of Brittany sent to put down the "stamped-paper rebellion" of 1675. Works on eighteenth-century army intendancies include studies of Jean Moreau de Séchelles in the War of the Austrian Succession, François-Marie de Gayot during the Seven Years' War, and Joseph Benoist de Tarlé in the army of the count de Rochambeau serving in America during the War of Independence.[2]

---

1. Louis André, *Michel Le Tellier et l'organisation de l'armée monarchique* (Paris, 1906), pp. 628–40 (hereafter cited as *Michel Le Tellier et l'armée*); Louis André, *Michel Le Tellier et Louvois* (Paris, 1942), pp. 414–27.

2. Charles Schmidt, "Le Rôle et les attributions d'un 'intendant des finances' aux armées, Sublet de Noyers, de 1632 à 1636," *Revue d'histoire moderne et contemporaine* 2 (1900–1901): 156–75; Narcisse-Léonard Caron, *Michel Le Tellier: Son Administration comme intendant d'armée en Piémont, 1640–1643* (Paris and Nantes, 1880); Séverin Canal, "Essai sur Auguste-Robert de Pomereu, intendant

Military historians have in general, however, shown an amazing neglect of the army intendancies. Even experts living in the ancien régime, such as Honorat de Meynier, Louis de Gaya, Father Gabriel Daniel, the marquis de Feuquières, and François de Chennevières, paid scant attention to this subject. Honorat de Meynier wrote a treatise on military life, *Reigles, sentences et maximes de l'art militaire* (1617), but he made no mention of the army intendant, although he did comment about the military commissioners, the *commissaires des guerres*. Louis de Gaya's *Art de la guerre* (1677) remarked that since the author's design was to examine all those employed in the armies of his day, he could not forget the intendants, commissioners, and treasurers because they deserved a word in passing, but he confined himself to a few sentences on each. Much more astonishing was the treatment—or lack of treatment—by Father Gabriel Daniel in his famed two-volume study of the French army, *Histoire de la milice françoise* (1721); he did not devote any section to the army intendant, although he lavished copious detail on the regular army officers. The marquis de Feuquières's *Mémoires* (revised London edition of 1736), which was really an account of the principles of military operation, dealt with the army intendant in a single page, remarking: "When we have spoken of the function of the intendant, we have said almost everything that regards the *commissaire des guerres* in the army." The eighteenth-century historian François de Chennevières, spent more time on the subject in his *Détails militaires* (1750), but his comments were more like an instruction manual than administrative history. For example, he remarked that an intendant needed at least three secretaries—four in a large army —to assist him in his responsibilities.[3]

d'armée en Bretagne (1675–1676)," *Annales de Bretagne* 24 (1908–9): 497–513. For the eighteenth-century army intendancies, see Jean des Cilleuls, "Un Grand Intendant d'armée: Moreau de Séchelles (1690–1760)," *Revue de l'intendance militaire*, no. 26 (1953), pp. 39–79; Charles-Nicolas Dublanchy, *Une Intendance d'armée au XVIIIe siècle: Etude sur les services administratifs à l'armée de Soubise pendant la guerre de sept ans* (Paris, [1905]); Jean des Cilleuls, "Le Service de l'intendance à l'armée de Rochambeau," *Revue historique de l'armée*, numéro spécial consacré à la fraternité d'armes franco-américaine (1957), pp. 43–61.

3. Honorat de Meynier, *Les Reigles, sentences et maximes de l'art militaire, et les remarques du sieur de Meynier sur le devoir des simples soldats, et de leurs supérieurs* (Paris, 1617), p. 41; Louis de Gaya, *L'Art de la guerre et la manière dont on la fait à present* (Paris, 1677), pp. 29–30; Gabriel Daniel, *Histoire de la milice françoise, et des changemens qui s'y font faits depuis l'établissement de la monarchie dans les Gaules jusqu'à la fin du règne de Louis le Grand*, 2 vols. (Paris, 1721); Antoine de Pas, marquis de Feuquières, *Mémoires de M. le marquis de Feuquiere, lieutenant général des armées du roi; contenans ses maximes sur la guerre, et l'application des exemples aux maximes*, new ed., rev. and cor. (London, 1736), pp. 58–59. François de Chennevières, *Détails militaires, dont la connoissance est nécessaire à tous les officiers, et principalement aux commissaires des guerres*, 6 vols. (Paris, 1750), 4:165.

At the beginning of the nineteenth century, Xavier Audouin referred briefly to the army intendants in his multivolumed study of military administration, *Histoire de l'administration de la guerre* (1811), but he did not deal systematically with this office. He did, nevertheless, spend considerable energy tracing the evolution of the *commissaires des guerres*. From the powers he attributed to these commissioners, he may have been confusing them with the army intendants, some of whom were drawn from the commissioners' ranks. The well-known military historian François Sicard devoted only two pages to the office of intendant in a four-volume study, *Histoire des institutions militaires des Français* (1834), repeating such common information as the fact that the army intendants were chosen from the ranks of the provincial intendants and the military commissioners. He was mistaken, however, in his belief that the title of intendant of the army was not used before 1635. A later historian, Edgard Boutaric, penned no more than a paragraph on the army intendant in his *Institutions militaires* (1863), while one of his contemporaries, Louis-Maximilien-Modest Chassignet, author of *Essai historique sur les institutions militaires* (1869), spent a couple of pages discussing intendants without distinguishing between those in the province and in the army. Chassignet's remarks are so general that they border on the commonplace. For example, ignoring the instrumental role of Le Tellier, he noted that Louvois followed Richelieu's example of placing an intendant next to a general, both as a chief administrator and as a delegate of ministerial power.[4]

One can only speculate why these standard institutional histories ignored the office of army intendant. Most of the seventeenth- and eighteenth-century writers were more interested in composing manuals of military conduct than administrative history. Their interest was more in "men of the sword" than in "men of the bureau," and their volumes are full of details about the operation of sieges and military maneuvers. Nineteenth-century historians took a more encyclopedic view, trying to touch briefly on all aspects of military life rather than to treat the army intendancy in detail. Then, too, because the office of intendant of the army disappeared with the fall of the ancien régime, most of these historians ignored it, for they were concerned with the

4. François-Xavier Audouin, *Histoire de l'administration de la guerre*, 4 vols. (Paris, 1811), 2:24-31; François Sicard, *Histoire des institutions militaires des Français*, 4 vols. and atlas (Paris, 1834), 1:416-17; Edgard Boutaric, *Institutions militaires de la France avant les armées permanentes* (Paris, 1863), pp. 379-80; Louis-Maximilien-Modest Chassignet, *Essai historique sur les institutions militaires, ou la formation, l'organisation et l'administration des armées en France, depuis les temps les plus reculés jusqu'en 1789* (Paris, 1869), pp. 287-89, 304-5. In Chassignet's favor, however, it should be mentioned that historians have recognized Le Tellier's importance only after the research of Louis André, cited above in **n. 1.**

evolution of the military institutions of their own period. Finally, many historians of the present day have avoided military history, either in distaste for the reality of war, or as is so often the case in the United States, in not wanting to be tagged a "military historian," which has generally been equated with an antiquarian mind occupied with cataloging battles and troop positions. But the role of army intendants is a legitimate concern of present historians of the seventeenth century. The army intendancy was a civilian office, an extension of the monarchy's authority, and in an age of bureaucracy concerned with the problem of how executive decisions are carried out, it is a proper area of investigation for modern historians.

This book focuses on these army intendants and tries to answer certain questions: what was the function of an army intendant—that is, what did he do? what type of men occupied this position? and what role did they play in the complex struggle of royal authority versus the particularism of privileged groups in the seventeenth century? Also this work has tried to differentiate clearly the army intendant from the provincial intendant, a distinction that is too frequently overlooked.

Additionally, this study examines an exceptional type of military intendant closely associated with those in the army, the so-called *intendants des contributions* ("intendants of conquered territories" or *intendants des levées*), usually staffed by the same men who occupied the regular army intendancies. Not only did these special intendancies include the same personnel but they had other similarities: since a conquered territory had an occupying army, these intendants exercised many of the functions of the army intendants; because these intendancies had no permanent boundaries, their jurisdiction varied according to the fortunes of war or changes in governmental policy; and, finally, because of peculiarities in the French administrative system, both the army intendants and these frontier intendants were responsible to the same official, the secretary of state for war.

The various secretaries of state retained a dual jurisdiction. Not only did some of them have a particular department like war or foreign affairs, but they also handled the incoming correspondence and outgoing royal replies to royal officials in certain geographical divisions of France. The war secretary usually received the frontier regions as part of his department because the correspondence obviously related to military problems. It was also normal for a secretary of state to obtain the appointment of his own clients, generally bound to him by lines of patronage and parentage and characterized in seventeenth-century French by the term *créature*. Consequently it was normal for the army intendants and the intendants of a conquered territory to have the same

master. Because these officials are so closely related to the army intendancy, I have included material on them.[5]

There are limitations in this study. First, it is narrow in scope; it does
not attempt to deal with the entire history of the army intendancy
throughout the ancien régime. Time and circumstance restricted my
attention to the years between 1630 and 1670. A scholar with a grant
for a year or more of research simply does not have the time to go
through all of the vast collections in the Bibliothèque Nationale, the
Archives Nationales, the Archives des Affaires Etrangères, and various
other libraries, trying to examine all of the documents relating to this
problem. This type of research can be done only by French historians
like Pierre Goubert, or more recently Michel Antoine, who have had
years to work their way through the complexity of archival holdings.
However, I investigated all of the above archives, and made several ex-
ploratory trips to the Bibliothèque de l'Arsenal and the Bibliothèque
Mazarine. But the paucity of readily identifiable material meant that
most of the time spent was concentrated at the war archives at Vincennes, run by the *Service historique, Etat-major de l'armée de Terre*,
commonly known as the Archives de la Guerre. These archives preserve most of the surviving military records of the ancien régime.[6]

Even then, the limits of time prevented me from examining more
than about two hundred manuscript volumes. Consequently I limited
myself to the period from 1630, the date of the earliest documents, to
1670, the date of the occupation of Lorraine before the beginning of
the second of Louis XIV's wars, the Dutch War. These dates are not as
arbitrary as they might seem. Scholars recognize that the 1630s and
1640s were the crucial period in the formation of the intendant, both
provincial and military. Particularly in military history, historians have
discovered, thanks to the work of Louis André, that Louvois was not
the great innovator that nineteenth-century historians such as Camille

5. For a brief discussion of the workings of the office of secretary of state, see
Orest A. Ranum, *Richelieu and the Councillors of Louis XIII: A Study of the
Secretaries of State and Superintendents of Finance in the Ministry of Richelieu,
1635–1642* (Oxford, 1963). Ranum has probably done more than any other American historian to popularize the term "creature"; see particularly his comments,
pp. 28–29. But the French term "fidèle" is also appropriate; cf. Roland Mousnier's
remarks in *Peasant Uprisings in Seventeenth-Century France, Russia, and China*,
trans. Brian Pearce (New York and Evanston, 1970), p. 25.

6. Pierre Goubert spent twelve years (1944–56) preparing his magisterial study
of the economy and society of Beauvaisis, *Beauvais et le Beauvaisis de 1600 à 1730,
contribution à l'histoire sociale de la France du XVIIᵉ siècle* (Paris, 1960), recently
republished in an abridged paperback edition by Flammarion as *Cent Mille
Provinciaux au XVIIᵉ siècle* (Paris, 1968). More recently, Michel Antoine has
published the results of his long years of research in the Archives Nationales, as
Le Conseil du roi sous le règne de Louis XV (Paris, 1970).

Rousset once believed. In large part, Louvois simply expanded and clarified the more original work of his father, Le Tellier, and the preliminary efforts of Le Tellier's predecessor as secretary of war, Sublet de Noyers.[7]

While it is a historical axiom that very little can be safely said for very long since more recent studies are always modifying and correcting older ideas, it is apparent that the pattern of the army intendancy throughout the seventeenth century had been firmly fixed by Louvois's active participation in the war department in the 1660s. I had time to examine some of the military correspondence of the Dutch War and found that the form of the army intendancies fitted the form already traced by Le Tellier. The few studies of eighteenth-century intendants cited in preceding paragraphs seem to indicate that the pattern of the army intendancy did not change substantially in that century either, except in complexity and detail. Of course much more work remains to be done before historians can produce a clear picture of the army intendancy throughout the course of the ancien régime, but it is hoped that this study will be a fruitful beginning.

A second limitation lies in the nature of the documents themselves. Three types of manuscripts in the A [1] series of the war archives were utilized: 1) copies of commissions, inventory lists, and rough drafts of replies kept by the secretaries, 2) bound copies of correspondence between the secretary of state for war and various intendants, generals, and other officials, and 3) a number of original letters bound in volumes according to date. The materials relating the 1630s are the most sketchy, since this was the period before the Le Tellier family began the systematic retention of records from the war bureau. The materials relating to the 1630s are primarily copies of commissions, often only first drafts or, at most, corrected parchment versions that give a partial picture of the army intendant in this period, a portrait that must be enlivened by various printed collections of correspondence. The amount of documentation in the A[1] series increased significantly during the period before the Fronde, but most of the material consists of bound copies of selected letters from and to Le Tellier. Even these copies dwindle for the 1650s, and it is very difficult to comment on the army intendant

7. Camille Rousset, *Histoire de Louvois et de son administration politique et militaire*, 4 vols. (Paris, 1862–63), ignored Le Tellier's efforts, remarking in vol. 1, p. 173: "Les questions d'administration proprement dite ne le touchaient guère, non plus que la gloire des réformes; il évitait ou il tournait les difficultés, sans les resoudre," and speaking of the "insouciance de Le Tellier." In stressing Louvois's role to the detriment of his predecessors, Rousset only followed in the paths of previous historians; see Louis André's introduction to *Michel Le Tellier et Louvois*, pp. 9–10. Regarding the Le Tellier family as continuators of Richelieu's work, see André, *Michel Le Tellier et l'armée*, pp. 361, 415–16.

during this period. In the 1660s, however, the amount of documentation increased once more, and by the 1670s the researcher confronts volumes of material. Obviously the selective retention of some but not all of the original documents makes the task more difficult for a historian. He might fiind one or two letters relating to an incident, but no record of its outcome—an event that frequently is frustrating.

Special methodological problems exist. In the chapter on the army intendants and their world, my procedure was simple: I selected two dates, 1630, the beginning of the records in the war archives, and 1691, the death of Louvois. Although I did not examine all the material relating to Louvois's administration, it was important to give as complete a picture as possible of the personnel of the army intendancies, because many of the army intendants who served under Louvois were part of his father's patronage system. As Louis André discovered a little more than three decades ago, it is difficult to separate the ministries of the two. Therefore I decided to include as many of Louvois's intendants as possible in order to depict the entire family patronage system. The list of intendants includes those appointed after 1670, the date at which this study stops, but the inclusion of these names in no way changes my conclusions.

Once the 1630 to 1691 time span had been established, I compiled a list of all the army intendants who had served between these two dates. A man could have served in numerous military intendancies or in only one, and then as a substitute. There was only one criterion—an official had to serve as an army intendant sometime during his long and varied career. The result was a broad, heterogeneous group that I believe accurately reflects the evolution of the army intendancy. Naturally this list cannot be regarded as complete, because copies of their official commissions were not always preserved. Even with the commission, evidence is sometimes lacking whether the appointee actually served. I tried to be as rigorously selective as possible, including only those army intendants for which I saw positive evidence, either in the form of commissions or in other documents in the war archives. Only if some published work, such as Séverin Canal's article on Auguste-Robert de Pomereu, had clearly established proof of an army intendancy, was the official included in this list.[8]

These precautions are important, because they conform—to the best of my ability—to the criteria set up by Edmond Esmonin for the establishment of a complete list of intendants. Although Esmonin was thinking of provincial intendancies, he stipulated that the list should, first, carefully define the word "intendant"; second, it should establish

8. See Canal, "Essai sur Auguste-Robert de Pomereu."

the exact name of each intendant; third, the list should include the date of the commission and the date of the intendant's arrival and departure. In reality, Esmonin admitted that the difficulties in finding this information might be insurmountable.[9]

I tried, with some success, to meet these conditions. I defined the army intendant to include only those officials who bore some variation of the title intendant of justice, intendant of finance, or intendant of justice, police, and finance in the army. The listing excluded various intendants of frontier places such as Casal, intendants of conquered territory, and vague officials called *intendants des vivres*. No provincial intendant was included unless his commission specified that he was to serve as an intendant of an army as well as a province. For this reason, the list of intendants in the appendix gives the exact French terminology of the document that confirms the intendancy.

I tried to meet Esmonin's second condition: the exact name of the intendant. But as Esmonin remarked, this is often very difficult, because an individual might be known by the title of his lands rather than his surname. Even with a last name, it is sometimes impossible to distinguish various members of the family, particularly a father and his son. Unless a positive identification could be made, I did not attempt to give the intendant's full name. For example, Jeannin, Le Goux de la Berchère, and the sieurs de la Court and Croix, and Ollier remain unidentified. The Talon family presented a similar problem, for various members of this family are known to have served as army intendants, but it was impossible for me to decipher their individual names for each intendancy.

Finally, Esmonin's third condition required specific dates of service. This, too, was a difficult taks. Many of the documents necessary to establish such a chronology are missing; for example, not all commissions give the exact date by day, month, and year—some, especially first draft versions, provide only the year. In other cases even the year is questionable. Often the correspondence that would furnish the date of an intendant's arrival in the army is incomplete, and most of the letters of recall are also missing. Only the absence of an intendant in the correspondence provides a clue, and that an imprecise one, as to his departure. As a result, I decided to simplify matters by listing the intendants only by their year of appointment.

Once the list of intendants had been set up and their names identified,

9. Edmond Esmonin, "Note sur la publication d'une liste des intendants des origines à 1789," *Etudes sur la France des XVIIe et XVIIIe siècles*, Publication de la Faculté des lettres et sciences humaines, Université de Grenoble (Paris, 1964), pp. 19–23. This article was originally published in the *Bulletin* of the Société d'histoire moderne in 1910.

I followed the method used by Roland Mousnier and François Bluche in their attempts to analyze the social position of French magistrates: the identification of the profession of the intendant's first known ancestor—with the exception of the father and grandfather—the grandfather's and father's own occupations as measured by their most important office, the occupation of the intendant's father-in-law, the occupation of the intendant before assuming the intendancy, his highest office after service as an army intendant, and, finally, the occupation of the intendant's male children—the seventeenth century being a bastion of male chauvinism.[10]

Here, too, the possibility of error was great. I used the genealogical records, collectively called the *cabinet des titres*, preserved in the manuscript room of the Bibliothèque Nationale, assisted by the genealogical information in the specialized studies of various twentieth-century social historians such as Mousnier, Bluche, and Frondeville.[11] Despite the judicious use of these manuals and the manuscript sources, it was almost impossible not to make some mistakes. In part, this is because the genealogical records in the *cabinet des titres* were sometimes deliberately falsified in order to maintain a family's prestige. Since ennoblement was a major social goal for many *roturiers*, it is obvious that the more ambitious and unscrupulous submitted false information to justify their claims. The royal genealogists of the ancien régime, the d'Hozier family, tried to correct this hodgepodge of misinformation and forgetfulness, but they often found themselves baffled—as indicated by their sometimes caustic remarks in the manuscript collection. Fortunately the historian can rely upon the d'Hozier family as his first defense, followed by the secondary research of twentieth-century scholars. With their assistance, an approximate picture of an intendant's social position can be obtained, but individual errors are unavoidable.

Seventeenth-century documents provide other difficulties. The colloquial expressions, the handwriting, the unfamiliar abbreviations, the contemporary allusions, and the tiny details of military practice

---

10. Roland Mousnier, ed., *Lettres et mémoires adressés au chancelier Séguier (1633-1649)*, Publication de la Faculté des lettres et sciences humaines de Paris, series "Textes et documents," tome 6 (Paris, 1964) (hereafter cited as Mousnier, *Lettres au Séguier*); François Bluche, *Les Magistrats du Parlement de Paris au XVIIIe siècle (1715-1771)*, Annales littéraires de l'Université de Besançon, vol. 35 (Paris, 1960).

11. Henri Frondeville has published a series of genealogical dictionaries on Normand parlementarians based upon the "manuscrit Bigot" in the Bibliothèque de Rouen. I found *Les Présidents du Parlement de Normandie (1499-1790)* (Paris and Rouen, 1953) and *Les Conseillers du Parlement de Normandie au seizième siècle 1499-1594)* (Paris and Rouen, 1960), published by the Société de l'histoire de Normandie, the most helpful of the four volumes that have appeared to date.

vaguely referred to, but never clarified, pose immense problems of translation and interpretation. One simply cannot accumulate in a few short years the depth of knowledge needed to solve all the problems. Even hours spent poring over secondary accounts of etymological dictionaries are not always elucidating. Accordingly, one is tempted to quote the caution of Orest Ranum in the preface to his study of Louis XIII's councilors: "The room for errors was therefore very great, and even after numerous verifications, it was impossible to remove them all." I have been constantly aware of the dangerous terrain, but it is impossible not to have made mistakes. For these I bear full responsibility.[12]

Another difficulty appeared in my effort to incorporate portions of the documents into the text. The translation was an attempt to put seventeenth-century French into contemporary English without losing the original flavor. The result was often a tension between too literal and too loose an interpretation. This same problem is evident in translating titles. Generally this study has attempted to put many French terms into English equivalents, while recognizing that there really is no exact parallel between *maréchal* and marshal, *généralité* and generality, *premier président de parlement* and first president, *prévôté* and provostship, *subdélégué* and subdelegate, *châtellenie* and castellany, *secrétaire d'état de guerre* and war secretary.[13]

Similar problems arose in transcribing geographical names. Many of the places mentioned in the text have various spellings; for example, Ghent (English), Gand (French), and Gent (Belgian). The most frustration occurred in eastern European place-names. The city Oedembourg mentioned in Louis Robert's letters is actually the equivalent of the German Ödemburg, which is present-day Sopron in western Hungary. The same is true for the town referred to as Marbourg in Robert's correspondence; the German equivalent is Marburg, but it is now known as Maribor in Yugoslavia. After some indecision, I decided to put the spelling of such places in modern French or German, except where there is an obvious English equivalent such as Dunkirk for the French Dunquerque.

When it comes to proper names, however, I have done the exact opposite, leaving the proper names of the intendants in the most common seventeenth-century form; for example, Champlastreux instead of the present Champlâtreux. Even then, there were variations. Docu-

---

12. Ranum, *Richelieu and the Councillors*, p. v.

13. Many other terms, particularly the names of seventeenth-century offices, have been kept in the original French because there is no real modern equivalent. For example, the term *commissaire des guerres* is used throughout the text because I felt that "commissioner of war" was not a suitable translation.

ments exist spelling Jean Balthazar's name as Baltazard and Baltazar. There really is no rule in this study except consistency.

In the footnotes, I have retained the original French including the irregular spelling and lack of accents marks. Seventeenth-century manuscripts were erratic in these matters; sometimes the same word in a paragraph would have an accent mark in one place, but not in another. I have not attempted to change the exact transcription of these passages, particularly in the appendix, except to reduce the haphazard capitalization to lowercase letters. No attempt was made to correct the punctuation—which was often lacking.[14]

Finally, in the course of any work, it is impossible to thank all the people who assisted in the effort. I wish to thank particularly the Franco-American Educational Exchange Commission for a Fulbright travel grant, the Alliance Française de New York for a fellowship covering living expenses for my stay in France (September 1967 to December 1968), and Dean Francis M. Boddy and the University of Minnesota Graduate School for aid in making the publication of this work possible. While in France, I received the help of numerous archivists, especially at the Bibliothèque Nationale and the war archives in Vincennes. At Vincennes, the staff of the *Service historique* befriended this young American and introduced me to the complexity of the war archives. This manuscript has benefited from the suggestions of Professors Paul W. Bamford of the University of Minnesota and Phillip N. Bebb of Ohio University. The latter's never-exhausted patience and long hours of criticism have undoubtedly improved this study. Most important, I wish to thank my adviser, Professor John B. Wolf, without whose encouragement, friendship, and advice this essay would not have been possible. Professor Wolf has probably done more than any other American scholar to interest young historians in the study of Ludovician France, and his enthusiasm has been infectious to me. Finally, I wish to thank Madame Schil for her kindness and Mrs. Linda J. Pedigo for typing the manuscript.

14. At the time of writing I had not had a chance to examine two works that are helpful to the topic: Adriana Petracchi, *Intendenti e prefetti: L'Intendente provinciale nella Francia d'antico regime.* Vol. I, *1551–1648* (Milan, 1971), and Jean-Louis Bourgeon, *Les Colbert avant Colbert: Destin d'une famille marchande,* Publication de la Sorbonne, collection n. s. Recherches, 6 (Paris, 1973).

SERVANTS OF THE SWORD

I

Introduction:
Origins of the Army Intendants

The past does not yield its treasure easily—the origins of the office of army intendant are difficult to decipher. The very term *intendant* provides the historian with few clues. Originally the word meant someone who directed another person's affairs. The well-to-do, for example, hired intendants to manage the household and properties, and during the course of time these agents became the butt of jokes: La Fontaine defined an intendant as one who fished in troubled waters, growing rich as his master grew poorer; Archbishop Fénelon poked fun at women who scrimped on candles but allowed their intendants to take them for everything they owned.[1]

Gradually the term came to refer to a person who directed some aspect of government administration, for, like the rich, the state employed supervisors to manage its affairs. Numerous officials during the seventeenth century bore some variation of the title of intendant: there were *intendants des batiments du roi*, who tended the royal palaces, gardens, and parks under the supervision of the *surintendant des batiments*, their director in chief; *intendants des eaux et fontaines de France*, who oversaw the maintenance of streams, canals, and aqueducts; and *intendants de la marine*, who policed France's seaports. More important officials, the *intendants des finances*, varying from two to twelve in number, looked for fraud and corruption in the handling of the government's finances, while their superior, the surintendant, stood at the peak of the fiscal administration until the abolition of this office in 1661 and its replacement by that of a controller general. Finally, there were the well-known *intendants de la justice, police et*

1. Littré's *Dictionnaire de la langue française*, s.v. "intendant"; Robert's *Dictionnaire alphabétique et analogique de la langue française*, s.v. "intendant."

finances, employed in both the provinces and the armies, who gradually gathered all matters of local and military administration into their hands.[2]

These latter two types, the provincial intendants and the intendants of the army, must not be confused. While they had similar origins, the two intendancies developed into quite different institutions. The army intendants' mission remained temporary, limited to a special circumstance—the military campaign—while the provincial intendancy became a permanent fixture in the provinces, never vacant for any length of time. This permanency contributed to the growth of the provincial intendants' power until they formed a network of officials throughout the country and constituted the heart of the royal administration in the interior of France. They claimed extensive judicial powers on a level with the parlements, such as reviewing court procedure, overturning judgments, and transferring legal proceedings from one court to another. Sometimes, with special commission from the crown, they even called together other legal officials and constituted a court without appeal except to the king. They presided over the apportionment of the *tailles,* held the dominant voice in the bureaus of the treasurers of France—the principal fiscal officials in the provinces who were in charge of the royal domains and taxation—verified the accounts of the local receivers, and generally supervised tax collection. They organized and controlled the urban police and the royal constabulary, saw to the safeguarding of highways from brigands, inspected the repair of roads and bridges, policed the markets and fairs, reported on the economic life of the region, and, in general, represented the central authority of the crown.[3]

Provincial intendants, however, performed many military functions common to the army intendancy: recruitment, disciplining troops, payment, fiscal accounting, provisioning, canteen management, lodging, regulation of winter quarters, supervision of fortifications, and hospital administration. One is tempted to use a simple rule of thumb

2. Marion's *Dictionnaire des institutions de la France aux XVIIe et XVIIIe siècles,* 1969 reprint of 1923 ed., s.v. "intendants des batiments du roi," "intendants des eaux et fontaines de France," "intendants de la marine," and "intendants." For the "intendants des finances," see Julian Dent's helpful article, "An Aspect of the Crisis of the Seventeenth Century: The Collapse of the Financial Administration of the French Monarchy (1653–61)," *Economic History Review,* 2d ser., 20 (1967): 241–56. See also Claude Aboucaya's study, *Les Intendants de la marine sous l'ancien régime* (Paris, 1958).

3. Robert Mandrou, *La France aux XVIIe et XVIIIe siècles,* Nouvelle Clio, no. 33, rev. 2d ed. (Paris, 1970), p. 217. For further information on the provincial intendants, see the bibliographical article of Maurice Bordes, "Les Intendants de province aux XVIIe et XVIIIe siècles," *L'Information historique,* May–June 1968, pp. 107–21.

to delineate their separate spheres: if an activity involved a province, it was the responsibility of the provincial intendant; if it involved an army in the field, it was the army intendant's duty. In reality, such divisions are difficult to make, for there was often close cooperation between an army intendant and any number of provincial intendants, particularly those along the frontier.[4]

Recruitment was the one activity in which an intendant of the army rarely took part, for his instructions usually called for him to join an already assembled army at a set time and place. Provincial intendants, however, played an important role in the various methods of recruitment existing in the seventeenth century. They often negotiated agreements with local French officers to raise a company in return for royal subsidies—the usual method of recruitment—and they inspected these new companies to certify the full number of men and armament before authorizing payment to the captains. In troubled times like the Thirty Years' War and the Fronde, frontier intendants in regions such as Alsace contracted with foreign officers to bring their men into French service. Later in the seventeenth century, the provincial intendant had charge of the conscription of the militia, an annual contingent of men levied upon the local communities.[5]

Like the army intendants in the field, the provincial intendants were responsible for the troops stationed in their region. Regular reviews were conducted by military commissioners under the intendant's authority, the *commissaires des guerres*, to see that the troops were complete in number and properly equipped. Another part of an intendant's duty was the reduction of desertion and the practice of

4. For this and the subsequent paragraphs on the military duties of the provincial intendant, see Albert Croquez, *L'Intendance de la Flandre wallonne sous Louis XIV 1667–1708*) (Lille, 1912), pp. 285–98, and Charles Godard, *Les Pouvoirs des intendants sous Louis XIV particulièrement dans les pays d'élections de 1661 à 1715* (Paris, 1901), pp. 379–415. One should also consult the monumental works of André, who has a number of chapters on the reforms of Le Tellier, although he was concerned with the secretary's efforts and does not always clarify the role of the provincial or army intendants. See, in particular, chapters 5–10 on recruitment, pay, dress, lodging, supplies, and hospitals in *Michel Le Tellier et l'armée*, and for the years after 1660, chapters 8–10 in *Michel Le Tellier et Louvois*.

5. For example, see André, *Michel Le Tellier et l'armée*, pp. 242–43, for the role of Imbert, intendant of Languedoc, in contracting with French officers to raise companies for the army. For a sample of such an agreement during the Fronde, see the contract dated 17 May 1650, which was negotiated by Barthélemy Hervart, intendant of finance, and Philibert Baussan, intendant of the province of Alsace, with a German named Klüge, "general major" of the German artillery and colonel of a regiment of infantry, to keep his regiment in Mazarin's service, in the Archives de la Guerre, A¹ 122, fol. 169 (hereafter the initials B.N., A.A.E., and A.G. on manuscript citations refer to the Bibliothèque Nationale, Archives des Affaires Etrangères, and Archives de la Guerre; "r" and "v" after folio numbers refer to recto and verso pages).

passevolants, false musters of men who were not soldiers but valets or "hangers-on" dressed as soldiers in order to deceive the official count at reviews.[6]

The provincial intendant authorized the distribution of money to the troops at these reviews, although the actual payment was in the hands of assistants, *commis,* of the treasurer of war for that province. In addition, the intendant sent records of all military expenditures in the region to the secretary of state for war. Much of an intendant's activity, in fact, involved the dispatch of itemized accounts, and consequently the provincial intendant can be thought of as an important agent in the state's efforts to put order into its confused finances.

Provincial intendants, especially those on the frontier, also played an important role in supplying armies. Before the troops left on campaign, the crown ordered the intendants to stockpile necessary supplies—grain, hay, and straw—in magazines near the war theater. The intendant accumulated similar provisions for storage in depots, called *étapes,* to supply an army along its authorized route of march. At the beginning of the century, intendants levied goods in kind for the depots, but later the government changed this to a monetary sum deducted by the intendant from a community's share of the *taille,* while the intendant either contracted with a private entrepreneur or purchased the goods on the open market.[7]

Canteen management was another of the provincial intendant's preoccupations. It was customary for the government to set lower prices on beer and wine in special military bars called *cantines* to compensate the soldier for his low pay. Regulating the official price of drinks was no easy task, for it required consultation between the provincial intendant, the tax farmers who benefited from the taxes on alcohol, and the local magistrates. At the same time, it was also the intendant's responsibility to handle complaints about the behavior of drunken and rowdy troops, complaints that were all too common.[8]

Lodging, too, remained a problem for the provincial intendant. This had several aspects: troops were stationed in permanent garrisons, temporarily housed as they moved from one area to another, or

6. Obviously the reforms of Le Tellier and Louvois did not succeed in stamping out the abuses of desertion and *passevolants;* see, for example, the remarks of André Corvisier on these abuses in the eighteenth century, as well as the inaccuracy of official figures, in *L'Armée française de la fin du XVIIe siècle au ministère de Choiseul: Le Soldat* (Paris, 1964), 2:579–80, and all of chap. 3, pt. 4, on desertion.

7. On the whole question of *étapes,* see André, *Le Tellier et l'armée,* pp. 421–26.

8. For examples of incidents between soldiers and civilians, see Alain Lottin, *Vie et mentalité d'un Lillois sous Louis XIV* (Lille, 1968), pp. 171–75.

quartered in provinces in the winter season between campaigns. France had no barracks for its soldiers, except along the frontier, and most soldiers camped out in the open or resided with the local inhabitants. The secretary of state normally selected the routes for the passage of troops and the towns in which they were to be quartered, although the choice of the latter was sometimes left to the provincial governors and intendants.

Minute regulations prescribed the order for lodging troops over a long period of time. As soon as an army entered the province, the mayor and town councilors, accompanied by a *commissaire des guerres*, went from house to house drawing up lists of available accommodations. Then, in the presence of the town officials, the governor, and the intendant, the actual allotment took place. *Billets de logement*, written by clerks and signed by the intendant, ordered a homeowner to admit so many soldiers, generally two or three, to his dwelling. The government permitted few exemptions, except clerics, nobles, and prominent officials, from this obligation. But if there was disagreement over exemptions, the governor and the intendant had the authority to settle differences.[9]

Soldiers quartered among households had the right of *ustensile*, whereby the law required the owner to provide bed, linen, heat, candles, dishes, and various food supplies such as salt, vinegar, and sometimes wine. The *ustensile* soon became an excuse for troopers to extort money and goods from their hosts, although numerous ordinances denounced this evil. The state left the enforcement of these regulations to the governor and the intendant, which necessitated close cooperation between the two.[10]

Fortification was another concern of the provincial intendant, particularly along the frontier, for Louis XIV encircled France with a ring of strongholds to help prevent the threat of foreign invasion. Although these public works were carried out by royal engineers such as Vauban, the intendant made regular inspection tours and reported the results to the secretary. Neighboring towns met the cost according

9. For the whole problem of lodging, see André-Eugène Navereau, *Le Logement et les ustensiles des gens de guerre de 1439 à 1789* (Poitiers, 1924), particularly pp. 221–22 for examples of *billets de logement*. For the procedure of lodging, see Pierre-Claude de Guignard, *L'Ecole de Mars ou Mémoires instructifs sur toutes les parties qui composent le corps militaire en France, avec leurs origines, et les differentes maneuvres ausquelles elles sont employées* (Paris, 1725), 1:221; André, *Michel Le Tellier et l'armée*, pp. 375–76.

10. Regarding the close cooperation that usually existed between the intendants and the provincial governors, see Roland Mousnier's important article, "Note sur les rapports entre les gouverneurs de provinces et les intendants dans la première moitié du XVIIe siècle," *Revue historique* 228 (1962): 339–50.

to an agreement negotiated with the intendant, who also had the authority to authorize actual expenditure for the construction work.

Finally, intendants had to provide medical care for the soldiers stationed in their provinces. This could be a serious problem in frontier regions when many armies passed through their boundaries. In practice, the intendant had to contract with various nursing orders of monks and nuns or with entrepreneurs for the care of the invalid.

Naturally, the handling of all these details, plus the intendant's other responsibilities, was beyond the ability of any one man. For this reason, the government stationed a special type of *commissaire des guerres*, the *commissaires provinciaux*, in each province to handle the everyday matters of military administration. These commissioners, however, were directly under the control of the provincial intendant, who had to render an account of their activities to the secretary of state for war.[11]

Because both the intendants of the army and the intendants of the provinces shared some of the same military duties, and because both officials bore the common appellation of *intendant de la justice, police et finances*, it is easy to assume a common origin, but the proof of this origin is not certain. Despite years of research, historians do not know much more now than they did a hundred years ago about the origin of these institutions. Partly this has been due to an imbalance of effort; historians have argued about the beginnings of the provincial intendants but ignored the army intendants, although it was recognized that both sprang from the same source, the army camps. Consequently it is difficult to say who were the first intendants.[12]

It is known that the office was not created by Richelieu in 1635, as was formerly believed, nor, in fact, by any specific edict. Instead, it had a long evolutionary growth focusing about the monarchy's attempt to exercise its authority. In order to extend their influence in the countryside, the French kings had resorted to the practice of sending various types of inspectors into the provinces armed with commissions containing unlimited judicial and financial powers. These missions all had certain characteristics: the commissioners were non-

11. Godard, *Pouvoirs des intendants*, pp. 379–80. André, *Michel Le Tellier et l'armée*, pp. 620–21.

12. See Roland Mousnier's incisive comments on the current state of knowledge on intendants in his "Etat et commissaire: Recherches sur la création des intendants des provinces (1634–1648)," in *Forschungen zu Staat und Verfassung: Festgabe für Fritz Hartung*, ed. Richard Dietrich and Gerhard Oestreich (Berlin, 1958), pp. 325–26. See also Edmond Esmonin's comments, "Observations critiques sur le livre de M. Hanotaux: 'Origines de l'institution des intendants des provinces,'" *Etudes sur la France des XVIIe et XVIIIe siècles*, pp. 15–16, originally published in the *Bulletin* of the Société d'histoire moderne (1932–33).

native—arriving from Paris, their tour was restricted to a specific assignment, and because their commissions were deliberately worded to give them the full plenitude of power, they could make reforms in the face of strong opposition. Usually their judgments were without appeal, except to the king's council, so no local institution had the power to reverse their decisions.[13]

Historians have traced such royal commissioners back to the *enquêteurs royales* of Saint Louis's period, or even, as the seventeenth-century intendants themselves proudly claimed, back to the *missi dominici* of Charlemagne's era. Other historians have traced the origin of the intendants back to the circuits, the *chevauchées*, of a specific type of royal commissioner, the *maîtres des requêtes*, the masters of requests. These important government officials gathered background material and prepared dossiers for matters before the royal council in Paris, and during the fifteenth and sixteenth centuries various legislation prescribed tours in the provinces. For example, in 1553 the monarchy ordered an annual inspection trip for six masters of requests. They were to be provided with a special commission, their task was to investigate complaints regarding the administration of justice in the countryside, and they were to transmit their findings to the royal council and the local parlements.[14]

It is arguable, however, whether there is any direct link between these tours of duty and the missions of the intendants, although their position in the government made the masters of requests prime candidates for the position of intendant. The *chevauchées* were small in number and their personnel too few. Despite the repetition of legislation, they lapsed into an irregular basis in the troubled sixteenth century.[15]

The religious and political conflict in the sixteenth century, culminating in the accession of Henry IV to the throne, did require a large number of temporary and ubiquitous officials. Recent studies of these "precursors" or "pre-intendants" stress their almost infinite

13. Historians like Georges d'Avenel, *Richelieu et la monarchie absolue*, 2d ed. (Paris, 1895), 4:199–200, recognized that the intendants were anterior to Richelieu, despite the confusion over an edict of 1635 creating the office of "intendants general of finance." More recently, Roland Mousnier has declared that although this edict did not create the office of intendant, it was an important part of the crown's effort to twist tax control from the hereditary officeholders and put it into the hands of its own agents, thus marking a change in the 1630s from "inspecteur-reformateur" to administrator, a veritable new creation in Mousnier's eyes. Mousnier, "Etat et commissaire," pp. 327, 332–33.

14. Roger Doucet, *Les Institutions de la France au XVIe siècle* (Paris, 1948), 1:422–26.

15. See in particular Esmonin's criticism of the *chevauchée* theory, in "Critique du livre de M. Hanotaux," p. 14.

variety. They ranged from treasurers of France to royal lieutenants of the *grand voyer*, Sully, and included such officials as councilors of state, masters of requests, members of the sovereign law courts, and financial experts. The crown kept them busy reporting on financial, military, and administrative affairs. They oversaw work on roads and the rebuilding of canals and bridges ruined or neglected during the war, dismantled fortresses of unruly nobility, attended meetings of provincial estates seeking grants of money for the crown, investigated claims of tax misapportionment, negotiated with potential rebels, and, in general, made royal authority felt throughout France.[16]

A curious feature of their missions is the mixture of civilian and military activity. A limited study of the careers of two men, Raymond de Viçose and the sieur de Gastines, illustrates this combination. Raymond de Viçose was a naturalized Frenchman who had entered the service of Henry IV and accompanied him on his campaigns. In 1594 the king appointed him *intendant et contrôleur général des finances* and sent him to accompany Marshal de Matignon on his expedition to suppress sedition in the province of Guyenne. In the army Viçose proved indispensable to the marshal in financial matters. Two years later, he was at Rodez negotiating with the estates of Rouergue for grants of money and quarreling with the Parlement of Bordeaux. That same year, 1596, Viçose attended the assembly of notables at Rouen as a royal commissioner, after which he turned to fiscal tasks in Poitiers. A couple of years later, he traveled to Brittany to hire ships to support a military campaign against the duke of Mercoeur. Later in 1598, he was a royal agent at a Protestant synod, before making a tour of the Midi region to report on dissension in that area.

The activity of the sieur de Gastines is strikingly similar. In 1601 this *commissaire député pour la direction des finances* advised the king on the best use of cannon formerly employed in a campaign against Savoy. The following year he accompanied a French force that followed the course of Spanish troops in their passage along the frontier. He had orders to oversee the payment of troops as well as to assist the French commander. The subsequent year he was a royal agent at the meeting of the provincial estates of Languedoc.[17]

Given the number and the variety of these commissioners, it is impossible to distinguish the origin of the intendants in any one class of official. One can do little more than repeat the conclusion of another historian: "It would be in vain to search for a principle of classifica-

16. David Buisseret, "A Stage in the Development of the French *Intendants:* The Reign of Henri IV," *Historical Journal* 9 (1966): 38.
17. Ibid., pp. 28–31.

tion to distinguish the various categories of commissioners. Different titles were successively applied to the same person in the course of his mission, preventing us from attributing any importance to the incoherent possession of titles." [18]

The situation in the army was similar. There, too, the crown needed agents of inspection vis-à-vis the regular military corps. The top army commanders, the constable and the group of marshals beneath him, controlled all matters of command, inspection, discipline, and administration. They commissioned their own private secretaries and their fellow army officers as their inspectors, and these officials held a wide variety of titles, including lieutenant, deputy, clerk, *commis*, and commissioner. When the monarchy did intervene, it did not enter into the details of these agents' administration, but instead concentrated its attention upon the vital question of the control of troop payments. An important change came in the fourteenth century when the Estates General assumed emergency governmental powers during the captivity of King John II. It sent the first crown officials, twelve in number, into the army in 1357, "to watch the payment of wages to the men of war." In reality, the reform was not a radical departure; it had been in the air for some time, but now crisis made the practice necessary.

Succeeding monarchs, particularly Charles VII in the fifteenth century, multiplied the number of royal commissioners sent into the army as *commissaires deputés*, delegates of crown authority. Their missions were temporary and specific; the intention was not to replace the authority of the marshals, but to assist them in correcting abuses. The result was a wide variety of agents, of both the crown and the military hierarchy, each often bearing some form of the confusing title *commissaire des guerres*. Nevertheless, the activity of these commissioners expanded to include inspection of horses and armament, supervision of the payment of wages, conduct of reviews, search for *passevolants* and incomplete companies, accompaniment of troops from one locality to another, inspection to see that the orders of the sometimes distant commanders were obeyed, and supervision of the troops in winter quarters.

The sixteenth century was exceedingly important in the formation of this office. Francis I placed all the various commissioners, both civilian and military, under the jurisdiction of the chancellor, a civilian office representing royal authority. Letters of provision from the king installed these officials in office, and they again bore the general title of *commissaires des guerres*. These agents were not, how-

18. Doucet, *Institutions de la France*, 1:429.

ever, the most effective agents of royal authority. The constable and marshals kept various rights of nomination for themselves—even in the seventeenth century, every new marshal had the right of naming one *commissaire des guerres* upon his creation. Often these commissioners were exceedingly lax in their functions, tolerating various abuses like the *passevolants*, partly because the crown had little control over them. Although they were responsible to the chancellor, there was no immediate supervision, except by army officers, until the abolition of the office of constable in 1626 and the transfer of supreme military power to the hands of a civilian, the secretary of state for war. At the same time, the *commissaires des guerres* lost their original character as commissioners, although they retained the name in their title. This corps of officials was erected into an office, which meant that the positions were created by the government and bought and sold, according to the crown's fiscal needs, like any other office in France. Ownership of office made the officeholders virtually independent and unsuitable as the crown's chief agents in the army, but under the jurisdiction of the army intendant, the sheer number and technological background of the *commissaires* provided the assistance the intendants needed to make their authority felt in every area of military life.[19]

Although the intendants of the army did not originate among the corps of the *commissaires des guerres*, historians have accepted the idea that the first intendants appeared in military camps or at the side of generals during the troubled period of the sixteenth-century religious wars and the royal attempts at pacification of provinces. Little study, however, has been done on this subject, and the circumstances of the first intendants are unknown. A partial explanation might lie in a little-known manuscript in the Bibliothèque Nationale, attributed to Marshal Bassompierre around 1630. The document emphasizes that the great officers of the realm performed personal services of justice and finance when the king was present in the army. In their

19. This and the preceding paragraph are from Robert-Desiré Stiot, "Le Commissariat des guerres: Son organisation—son évolution—ses attributions," *Revue administrative*, no. 59 (1957), pp. 454-56; Jean Milot, "Evolution du corps des intendants militaires (des origines à 1882)," *Revue du Nord* 50 (1968): 382-86; D. Moulias, "Les Origines du corps de l'intendance," *Revue historique de l'armée*, no. 4 (1967), pp. 83-84. Despite the word "intendant" in the above titles, they do not deal with the creation of the office of army intendant; instead they are concerned with the formation of "l'intendance militaire," the special corps of the present-day French army dealing with the services of supply, sanitation, and medical care, essentially the combined offices of the seventeenth-century intendant and his *commissaires des guerres*. For the transformation and abolition of the office of intendant of the army during the French Revolution, see Jean Milot, "Du Commissaire des guerres à l'intendant militaire," *Revue historique de l'armée*, numéro spécial: L'Intendance militaire (1968), pp. 39-48.

absence, these services would be delegated to substitutes. The memoir states: "In the royal army, the chancellors or the keepers of the seals always perform the charge of intendant of justice, and when they are employed or diverted elsewhere, the king chooses one or two of his councilors of state of the long robe to substitute for them." They, in turn, employed various masters of requests to inform them of disorders and malversations as well as to conduct their trials. In like fashion, the memoir attributes the policing of the army to the grand provost, the administration of the hospital to the grand almoner, and "the handling and disposition of finances in the said royal army belongs to the surintendant, and, in his absence, they are administered by one of the intendants of finance." [20]

As late as the reigns of Henry IV and Louis XIII, the great officers of the realm still performed their ceremonial duties. Until Chancellor Bellièvre's departure for Marseilles to greet Henry's prospective bride, Maria de Medici, upon her arrival in France in 1600, he acted as intendant of the royal army engaged in the conquest of Savoy, and in his absence he entrusted the charge of intendant to the hands of Calignon, chancellor of Navarre, and President Jeannin. Fifteen years later, when the young Louis XIII traveled to the southern border to marry the infanta, Chancellor Sillery acted as intendant of justice for the troops accompanying the royal suite. Du Vair, keeper of the seals, served as army intendant at the battle of Ponts de Cé, and in the campaign of 1621, at the sieges of Saint Jean d'Angély and Clairac until his death early in August of that year. Constable de Luynes then assumed the keeping of the seals and, with it, the intendancy of the army, which he promptly delegated to De Vic and Caumartin. At Luynes's death on 15 December 1621, De Vic succeeded him as keeper of the seals. A later keeper of the seals, Marillac, exercised the intendance of the army at the siege of La Rochelle and later in Languedoc, while others, Bullion and Châteauneuf, held the intendance of justice in the army that forced the pass of Suze.[21]

According to the memoir attributed to Bassompierre, the two most important functions—justice and finance—were separate duties entrusted to the delegates of the chancellor and of the surintendant. Consequently there were at least two intendants instead of one in each

20. "En quoy consiste la charge et fonction d'intendant de justice et finances dans armée [*sic*]" bearing the notation "ce discours fait environ l'an 1630 [the original 1640 had been crossed out] par le maréchal de Bassompierre." B.N. *recueil Cangé* F168, boite K, tome 10, fols. 158r–63r. A second version, much easier to read, B.N. collection *cinq-cents Colbert* 499, fol. 99, is undoubtedly a copy that retains the date of 1640 and does not mention any author.

21. "En quoy consiste la charge d'intendant," *cinq-cents Colbert* 499, fol. 99.

army, the intendant of justice and the intendant of finance, for the two offices had not fused yet. This conclusion is reinforced by a "book of formulas" for commissions preserved in the Bibliothèque Nationale and dating from the reign of Henry IV. The formulary has blank commissions for both types of intendants. Those granted the "charge and intendance of justice"—the actual title "intendant of justice" was not used—were authorized to accompany a general in order to assist him at his councils, to report on various requests presented to him and give him advice on how to handle them, to listen to various complaints and try to reconcile differences, and to conduct the trial of malefactors found in the army. The commission concluded: "and generally do, in this charge and that which belongs to it, all that concerns the matter of justice, for the good of our service and the police of the said army." [22]

The commission for "the charge and intendance of our finances" in the army—again, the actual title of "intendant of finance" is not used—authorized the intendant to verify the statements presented to the military treasurers on the basis of which troop payments were made. He also had to inspect the account of the "general receipts of our finances," that is, of the expenditures made in the army, so he knew the current balance at all times. He had the responsibility of levying, in the general's name of course, various taxes and impositions necessary for the maintenance of the army, including the special aspects of artillery, victualing, and munitions supply. Although he had permission to use all the force he felt necessary for the prompt collection of the levies, he did have to render an account of these impositions to the government. Finally, no payment could be made in the army without the general's orders, countersigned—*contrerollés*—by the intendant. Obviously the intendant was the key fiscal accounting agent in the army.[23]

The connection between the intendants of justice and the chancellor's office has been commonly recognized, because many masters of requests, the chancellor's subordinates, held that commission; but until recently, few historians have understood that the intendancy of finance was a delegation of authority from the surintendant and his subordinates, the fiscal *intendants des finances*. This conclusion seems

22. "Commission pour l'intendance de la justice en une armée," B.N. *fonds français* 4,014, fols. 150v–51r.
23. "Commission pour l'intendance des finances en une armée," B.N. *fonds français* 4,014, fols. 151r–52r. Both the above copies are printed with slight variations from the original text in Gabriel Hanotaux, *Origines de l'institution des intendants des provinces d'après les documents inédits* (Paris, 1884), pp. 227–31. Since Hanotaux's transcriptions of documents were often hasty and error-ridden, I used the original manuscripts. See Edmond Esmonin's remarks about Hanotaux's deformed texts, in "Critique du livre de M. Hanotaux," p. 13.

inescapable if one reexamines the book of commissions for the reign of Henry IV, for the formulary specifically charged the army intendant of finance to act "with equal power and authority as would one of our intendants and councilors general of our finances, if he were present in person."[24]

There are various examples of intendants of finance in the top fiscal administration serving as intendants of the army in the early years of the seventeenth century. One of the most interesting cases is that of François Sublet de Noyers, who later became secretary of state for war. He served as intendant in the armies of the marshals d'Effiat and d'Estrées in 1632, and again from 1633 to 1635 in armies on the northern frontier and in Picardy. Prior to these missions, he had served as intendant of finance in Paris. Sublet de Noyers's biographer, Charles Schmidt, has noted this curious intermixture of civilian and military functions, for he confessed that Sublet was "an intendant of the army entrusted to the marshal d'Effiat to protect the Catholic princes of Germany. . . . It does not seem that he was, properly speaking, an *intendant d'armée*, but, indeed, an *intendant des finances* sent on special mission near the marshals, and his mission was, without doubt, to inform the cardinal [Richelieu] on what occurred in the armies."[25]

Intendants of finance were not the only important fiscal officials sent to the army as delegates of royal authority. Guillaume Bordeaux was a councilor of state with financial expertise who served as army intendant. He began his career in 1613, as a paymaster for one of the sovereign courts, the *chambre des comptes*. By 1633 he had obtained an appointment as secretary to the *conseil d'état et direction des finances*. That same year he received a commission to serve in the army of Lorraine commanded by the marshal de La Force, under the subcommand of the count de Brissac in Nancy. He continued to serve part of the following year in the army commanded by the marshals de La Force and Brézé. Only later, in 1649, did he himself become an intendant of finance in Paris.[26]

24. "Commission pour l'intendance des finances en une armée." B.N. *fonds français* 4,014, fols. 151r–52r. Séverin Canal, *Les Origines de l'intendance de Bretagne: Essai sur les relations de la Bretagne avec le pouvoir central* (Paris, 1911), pp. 27–37, is one of the few historians to have clearly recognized the separate institution of the intendants of justice and the intendants of finance in the armies in Brittany, and he thought that it was from the intendants of justice that the provincial intendants emerged.

25. Charles Schmidt, "Le Rôle et les attributions d'un 'intendant des finances' aux armées, Sublet de Noyers, de 1632 à 1636," *Revue d'histoire moderne et contemporaine* 2 (1900–1901): 157.

26. Commissions of Bordeaux as intendant of the army of Marshal de La Force in 1633, A.G. A¹ 14, fol. 87; as intendant with La Force and Brézé in 1634, A.G. A¹ 21, fol. 87; as intendant in the army and province of Champagne in 1635, A.G. A¹ 21, fol. 64. His biographical information is from Jean-Paul Charmeil, *Les*

Another example of an important fiscal official who served as army intendant is Gaspard Du Gué, treasurer general of France in the bureau of finances at Lyon. The case was highly unusual, because with the development of the provincial intendants, particularly between the formative years of 1635 and 1648, the intendants partially superseded the treasurers in the local bureaus of finance, causing great complaint from the corporation of treasurers general, who thought, with some justice, that the regular fiscal forms were being bypassed by royal action and their positions threatened. Du Gué, however, does not seem to have shared this prejudice. In 1629 he obtained a commission to share the position of "intendant of finance, supplies, munitions, and magazines" with Maupeou d'Ableges in the army of Marshal de La Force in the provinces of Burgundy and Bresse. The following year, he served by himself as "intendant of our finances, supplies, and magazines in the army of Italy commanded by the king in person, and, during the king's absence, by Cardinal Richelieu." [27]

It is obvious from the above examples that these army intendants were essentially important civilian officials who had little knowledge or expertise in the purely technical aspects of military life. And since many armies in the sixteenth and early seventeenth century were em-

Trésoriers de France à l'époque de la Fronde: Contribution à l'histoire de l'administration financière sous l'ancien régime (Paris, 1964), p. 260n137.

27. Commission of Du Gué as co-intendant with Gilles Maupeou d'Ableges, A.G. A¹ 13, fol. 149; as intendant of the army in Italy in 1630, A.G. A¹ 12, fol. 131. Prevost and Roman d'Amat, *Dictionnaire de biographie française*, s.v. "Du Gué de Bagnols (François)," attribute the intendancy of this army in Italy, as well as an intendancy in the army before La Rochelle, to Gaspard's son François. This seems to be an error; not only was François nineteen or twenty years of age at the time, but he did not possess the office of treasurer of France at Lyon, which Gaspard held from 1614 to 1651 and which office the two commissions clearly mention the intendant possessed. The genealogical records in the Bibliothèque Nationale do not identify any François de Gué as a child of Gaspard; instead they identify the François du Gué of the *Dictionnaire de biographie française*, and intendant at Lyon in the 1660s, as the son of François, a *maître ordinaire* in the *chambre des comptes* at Paris who was in turn a brother of the Gaspard who served as treasurer of France at Lyon. B.N. *dossiers bleus* 336, family "Gué, Du," fol. 11. This information on François is confirmed by B.N. *Biographies des maîtres des requêtes depuis 1575 jusqu'en 1722, suite des généalogies des maîtres des requêtes de François Blanchard*, in *fonds français* 14,018, p. 221rv. Despite the vast amount of information in the *Dictionnaire de biographie française*, it seems to be drawn mostly from printed secondary sources rather than the genealogical records in the manuscript room of the Bibliothèque Nationale. For an earlier case of an intendant of finance in an army in Brittany in 1593, François Myron, who held the position of *trésorier général des finances* in the same province, see Canal, *Origines de l'intendance de Bretagne*, p. 31. Roland Mousnier discusses the hatred of the treasurers of France for the provincial intendants, in his "Recherches sur les Syndicats d'*Officers* pendant la Fronde: Trésoriers généraux de France et élus dans la Révolution," *XVIIe siècle*, nos. 42–43 (1959), pp. 79–81.

ployed in the pacification of sedition in the countryside, it would be easy for an intendant to remain in a province after the army had left. In the words of the distinguished nineteenth-century historian Gabriel Hanotaux, there was an "insensible transition, the intendants of the army became intendants of provinces, at the same time conserving the title and functions of military intendant which they had at their origin." [28]

A study of early intendants in Languedoc shows this "insensible transition" from army to provincial intendant. In 1577 the king named Jean de Sade, sieur de Marsan, "to have the charge of justice near monsieur the marshal of Dampville" with authority to serve in the marshal's council of war to advise him on matters of military justice. At the same time, he was to preside at the *sièges présidiaux* in the area—in short, to function as a provincial intendant of justice. At Sade's side stood François Boyvin, sieur de Villars, who acted as "surintendant general of our finances" in the army of Languedoc. Boyvin concerned himself with "the good management and administration . . . [of the] large sums of money for the expenses, pay, and maintenance of the army." Both Boyvin and Sade served only for the campaign of 1577; they were temporary, not permanent, commissioners in the army or the province, but a precedent had been established for Languedoc.

In 1591 a sieur de Servière acted as "master of requests and intendant of justice near M. de Montmorency," governor of Languedoc, and after Servière's departure, his replacement, Claude de Convers, president of the *présidial* of Montpellier, bore the title of "intendant of justice near monseigneur the constable." Each year at the opening of the first session of the provincial estates of Languedoc, Convers found himself at the side of the governor or lieutenant governor and acted as spokesman for the crown. While Convers handled affairs of justice, another agent, Milles de Marion, treasurer of France at Montpellier, functioned as intendant of finance. He, too, in addition to advising the governor, attended sessions of the provincial estates as a royal commissioner—where he incurred the estates' rancor—and at the same time regulated the farming of the *gabelles*.[29]

The classic example of this transition from army intendant to provincial intendant is Charles Turquant, seigneur d'Aubeterre, councilor in the *cour des aides* in Paris, master of requests since 1585, and in 1595 intendant of justice in the army engaged in the pacification of Brittany under the command of the marshal de Saint Luc. Turquant was

28. Hanotaux, *Origines des intendants*, p. 46.
29. David Buisseret, "Les Précurseurs des intendants de Languedoc," *Annales du Midi*, n.s., no. 86 (1968), pp. 82–85.

truly a military intendant: he investigated complaints of fraud connected with a provincial treasurer of war, he attended the marshal's councils of war—providing sage advice against treating with the rebel duke of Mercoeur—and he kept the monarchy informed of everything that occurred in the army. His mission as intendant in the army lasted until 1598, when royal authority was restored in the province and the army disbanded, but instead of returning to Paris, Turquant remained in Brittany. During the years 1598 to 1610, he undertook a number of missions for the government. He attended sessions of the local parlement, forcing it to register unpopular edicts, such as that of 1599 granting the Huguenots certain rights in Brittany; he traveled to Nantes to settle differences between the inhabitants and the local governor. As royal spokesman at the sessions of the provincial estates, he negotiated that body's subsidies to the crown.[30]

It was natural that the intendant, the trusted agent from the royal council, should remain in a pacified province once the army left, for the monarchy had need of men of authority in both the army and the provinces in these difficult years. By the 1630s the intendants were a common phenomenon, but the institution still had no legal reality in the kingdom. Only the scrap of paper bearing the king's seal—their commission—gave the intendants their authority. Until the critical period of 1635–48, they remained only commissioners sent out on temporary missions as inspector-reformers. True, their numbers increased and their missions took a wider scope, but not until the crush of foreign war and internal economic difficulties were the provincial intendants transformed into a permanent and highly specialized institution.

Even then, Cardinal Richelieu, under whose administration they grew in number and influence, had no intention of establishing them on a permanent basis. In his *Testament politque* the words "intendant of justice" are found only once, and then in passing, referring to masters of requests sent out on temporary inspection tours. In fact, the passage suggested that it would serve the intendants' vanity more than the public welfare—"l'utilité du public"—if they were stationed in the main towns rather than forced to spend most of their time on circuits in the countryside. Nothing in the *Testament* suggests that the cardinal envisioned the intendants as more than special commissioners.[31]

30. Canal, *Origines de l'intendance de Bretagne*, pp. 38–51.
31. Mousnier, "Etat et commissaire," p. 327; Armand-Jean du Plessis, cardinal de Richelieu, *Testament politique*, ed. Louis André (Paris, 1947), pp. 246–47. For a brief discussion of the authenticity of the *Testament*, see the comments of Geoffrey Treasure, *Cardinal Richelieu and the Development of Absolutism* (London, 1972), p. 246n1.

Although the intendants made a successful transition from the army camps to the provinces, they retained a firm foothold in the military. Richelieu's ambitious foreign policy, and that of his successors, involved France in years of continuous warfare, making the army intendants indispensable. Nevertheless, the institution was not so specialized that the same official, the master of requests, the member of the royal council, or some important fiscal official, could not serve alternately in either the army or the province, now as an army intendant, then as a special commissioner or intendant in the provinces. Only at a later date, in the 1630s and 1640s, did the two types of intendancies become distinct.

II

The Army Intendants
and Their World

Any study of institutions requires an inquiry into the type of personnel employed and the relation to society at large. One's rank in French society during the ancien régime depended upon a number of factors, namely titles of nobility, antiquity of family, importance of one's office, marriage alliances, and, finally, that necessary ingredient, wealth. Naturally, those at the top of the social ladder possessed these factors to a large degree; great princes such as the Condé family held high titles of nobility, had impressive pedigrees, commanded important offices in the realm, contracted wealthy marriages, and owned extensive properties. These same factors provide a measuring stick with which to examine the social position of the officials who became army intendants.

Army intendants were drawn from a broad, heterogeneous group that reflected the evolution of the office of the army intendant. The classic example of an army intendant during the 1630s was an ex-councilor of parlement, a master of requests sent into the provinces to administer justice to an area wracked by rebellion. During the Thirty Years' War such men sometimes received additional commissions as army intendants to assist troop movements along the frontier or winter quartering in the provinces. In the period of the Fronde, when the position of the provincial intendant was officially abolished, army intendants acted extralegally in their place. The government recruited these army intendants from a wide variety of local and Parisian officials loyal to the crown, although the choice was sometimes taken out of the government's hands by powerful generals. With the return to calmer times in the 1650s and 1660s, Le Tellier constructed a unique team of relatives and clients, many of whom were ex–*commissaires des guerres*, to serve as control agents in the army. This pattern continued through the Dutch War, although in the later years of the Louvois administration army intendants seemed divided into two basic groups: long-term provincial intendants near the frontier who were familiar with border conditions and who could be recruited as army intendants for troops in their area, and ex–*commissaires des guerres* who had risen through the ranks to become army intendants in time of war and who sometimes made the successful transition to provincial intendants as a reward for their services.[1]

In brief, army intendants were drawn from a diverse background in the seventeenth century. They included officials from the sovereign courts, masters of requests, provincial intendants, regional officials, and *commissaires des guerres*. Despite its diversity, the group had one striking similarity: it consisted of a cadre of wealthy magistrates and administrative officials whose families had passed from the bourgeoisie into a service nobility, sometimes called the *noblesse de la plume* to set it off from the old-time aristocrats, the *noblesse d'épée*, as well as from the world of judges and other legal officials that comprised, strictly speaking, the *noblesse de robe*. This rise up the social ladder usually took several generations and followed a set pattern: bourgeois families progressively enriched themselves by wealthy marriages and the possession of fiscal offices, ennobled themselves through officeholding, and passed into the royal administration where this group of predominantly Parisian bureaucrats developed a certain esprit de corps, a mystique about *le service du roi*, or simply *le service*.[2]

One begins the study of this transition by an inquiry into the for-

1. The intendants of occupied places are not included in this social survey, because they were not, strictly speaking, intendants of justice, police, and finance in the army. Several, however, also served as army intendants—Carlier, Charuel, Clermont, Colbert du Terron, Gombault, Le Vayer, Robert, and Vauxtorte. It is therefore reasonable to conclude that the intendants of occupied places belonged to the same social milieu as the army intendants.

2. There has been considerable disagreement among historians about the nature of this social mobility. Some, like Roland Mousnier, have insisted that the legal fact of nobility cannot be equated with the social position of the nobility, for a deep chasm existed between the two. The old aristocratic families, well represented by the writings of Saint Simon, considered the "gentlemen of pen and ink" as bourgeoisie, despite the fact that they were legally aristocrats. Roland Mousnier, "Quelques Aspects de la fonction publique dans la société française du XVIIᵉ siècle," *XVIIᵉ siècle*, nos. 42–43 (1959), p. 5. Mousnier has stressed that this administrative nobility should be considered as a separate entity from the older aristocracy, although he does admit that they claimed social equality and sometimes even superiority to the *noblesse d'épée*. Cf. Roland Mousnier, *Etat et société en France aux XVIIᵉ et XVIIIᵉ siècles:* [Part 1] *Le Gouvernement et les corps*, Les Cours de Sorbonne (Paris, n.d.), pp. 24–27. Nevertheless, even Mousnier admits that administrative families attempted to transform themselves into the great families of the *noblesse d'épée*. He differs from other historians in stressing the difficulty, rather than the ease, of this transformation.

tunes of these officials, for whether society is based upon a division by estates or by social classes, it is wealth that facilitates the climb up the social ladder. Although the fortunes of most army intendants are unknown, there are random samples that provide some information on the subject. René de Voyer d'Argenson, sometime intendant of armies and provinces and later ambassador to Venice, left a fortune of 536,100 *livres*. Denis de Heere, who served many years as provincial intendant of Anjou, Touraine, and Dauphiné and as intendant of the army during the Fronde, had wealth in excess of 178,400 *livres tournois*. There are no exact figures for the fortune of Alexandre de Sève, intendant of the army in 1637 and later provincial intendant, but it was well over 600,000 *livres*. François Cazet de Vauxtorte, intendant of the province and army of Provence in 1640, had joint assets with his wife of 386,742 *livres* in 1642. Even an ex–*commissaire des guerres* such as Jacques Charuel left an estate of 350,000 *livres* after fifty years in royal service. Louis Robert likewise made a fortune in his career as intendant, but he lost much of it gambling. In one night he lost ten thousand *pistoles* to the duc de Lauzon—an incident that indicates the extent of his wealth.[3]

The composition and origin of these fortunes is unknown, but prosperity in the seventeenth century stemmed from a number of sources: wages and other emoluments from one's office, income from *rentes* and other invested wealth, familial inheritance, and land ownership. Usually the accumulation of such possessions took three or more generations of careful attention; only rarely was it the result of the self-made individual, the so-called new man.

From copies of "paychecks" in the war archives at Vincennes, the approximate wages of an army intendant are known. In the 1630s

3. Mousnier, *Lettres au Séguier*, 1:104–5, 118–19, 127, 159–60. For information on Charuel's fortune, including a summary of his will, see B.N. *dossiers bleus* 171, family "Charuel," fols. 2–13, 17–18. For an indication of Robert's fortune and gambling losses, see the notes of Gustave Servois in his edition of La Bruyère's writings, *Oeuvres de La Bruyère* (Paris, 1865–78), 1:505; one of La Bruyère's epigrams in his *Des Biens de fortune* concerned Robert's gambling.

The *livre tournois* was worth approximately one-quarter less than the Parisian *livre*.

Cf. François Bluche, *Les Magistrats du Parlement de Paris au XVIII° siècle (1715-1771)*, Annales littéraires de l'Université de Besançon, vol. 35 (Paris, 1960), pp. 149–53 (hereafter cited as *Magistrats du Parlement*), who mentions that eighteenth-century parlementary wealth was unhomogeneous in form and in amount, ranging from fortunes of 150,000 to 400,000 *livres* for moderate wealth to several million *livres* for the very wealthy. These are total fortunes, and Bluche notes that in the eighteenth century one could not lead a very grand life in Paris with an annual income of 15,000 *livres*. Higher officials, such as a president *à mortier* of Parlement, could not maintain the style of life expected from their office with an annual revenue of 40,000 *livres*.

army intendants received 600 *livres* per month as salary; after 1650 this sum increased to 1,000 *livres* per month. Provincial intendants received the same salary: 600 *livres* per month in the 1630s and 1,000 in the 1640s, underscoring the similarity between the two offices. As a yardstick it should be noted that *commissaires des guerres* received half this sum, or 300 *livres*, throughout the period of this study. In contrast, the duc de Duras, a marshal of France and a lieutenant general in the army of the dauphin in 1668, received 2,000 *livres* a month, although the other lieutenant generals serving in the army received only 1,000 a month.[4]

Like that of other officials, an army intendant's income did not depend upon the salary from his office. While eighteenth-century provincial intendants made between 15,000 and 70,000 *livres* a year from their salaries, royal pensions, and gifts from the crown and the province, army intendants also received gifts and pensions because of their office. For example, when the king named Jean Balthazar as intendant of the army of Lombardy in 1648, he authorized the *trésorier de l'épargne* to pay the intendant 3,000 *livres* "to aid him in making his

4. The "Ordonnance pour les appointments pendant les quatre premier mois de 1636 de Mr de Gobelin, intendant d'armée d'Allemagne," 20 April 1636, A.G. A^1 88, fol. 5, authorized the payment of 2,400 *livres* at the rate of 600 per month. See a similar "paycheck" for Louis Le Maistre de Bellejamme, intendant of the army of Picardy, for three months, or 1,800 *livres*, 11 November 1636, A.G. A^1 88, fol. 10. By 1650 Charles Brèthe de Clermont, "intend^t de la justice et finances en mon armée de Champagne," received the sum of 6,000 *livres* for six months "a raison de g# [1,000 *livres*] par mois," A.G. A^1 122, fol. 183. One thousand *livres* per month remained the salary throughout the seventeenth century. See the "Etat des appointements par mois des officiers generaux et majors qui servent dans l'armée de Monseigneur" for the year 1688, which mentions: "À l'intendant de ladite armée pour ses appointements pendant ledit mois la somme de mil livres," A.G. A^1 824, unnumbered pages at end of volume. There is also a pay receipt for Raymond Trobat, intendant of the army of Catalonia in 1690, for 2,000 *livres* "pour nos appointemens en lad. qualité pendant les mois de juin et juillet de la present campagne," B.N. *pièces originales* 2885, family "Trobat," fol. 1.

For provincial intendants, see the pay ordinance for Geoffrey Luillier d'Orgeval, intendant of Picardy, for 1636, dated 4 March 1637, which paid him at the rate of 600 *livres* per month, A.G. A^1 88, fol. 34. But by 1649 Charles Brèthe de Clermont received 1,000 *livres* a month as "intendant de la justice police et finances en nre pays d'Arthois," A.G. A^1 113, fol. 322. Cf. Albert Croquez, *L'Intendance de la Flandre wallonne sous Louis XIV (1667–1708)* (Lille, 1912), p. 49. For the pay of a *commissaire des guerres* and the duc de Duras, lieutenant general, see A.G. A^1 88, fol. 263, and A^1 824, unnumbered pages at end of volume, respectively. In comparison, see Bluche, *Magistrats du Parlement*, pp. 171–72, who puts the annual revenue from the office of an ordinary member of the Parlement of Paris (*conseiller aux enquêtes*) between 1,000 and 3,000 *livres*, certainly a modest amount, but this does not take into account the tremendous prestige of the office or the possibility of a sharp parlementarian using his office to gain additional profit.

voyage to the said country." This is considerably more than the amount of 1,200 *livres* that René de Voyer, sieur d'Argenson, received in compensation for the cost of his voyage to take up the army intendancy in Italy in 1637, so the travel allowance certainly kept up with inflation. Additionally, an intendant could expect various bonuses while on campaign. Ennemond Servien, intendant of the army of Italy in 1647, obtained an extra 300 *livres* per month as a salary increase. Louis Robert far exceeded this sum during the Dutch War, when he received a gratuity of 2,000 *écus* in 1673.[5]

Army intendants often received gifts of confiscated goods, particularly during the first half of the seventeenth century when many sporadic rebellions against royal authority led to a wholesale seizure of rebel property. In 1650 Jean de Choisy obtained a grant of "each and all of the goods, furniture, and property" that formerly belonged to a Michel d'Argouges, sieur de Bouville, appropriated by the crown for the crime of rebellion. Three years later, Ennemond Servien, intendant of the army of Italy, acquired the domains of the county of Lunel in Languedoc that had previously belonged to Count Antoine Gambera and forfeited to the crown by the treason of the above-mentioned count. In 1658, "in consideration of his long and faithful services," Talon, intendant of finance in the armies of Flanders and Luxembourg, gained possession of the movable property and real estate of one named Gremingue, inhabitant of Dixmude, whose goods were confiscated when he returned to his homeland upon declaration of war between France and Spain.[6]

Usually an intendant held other offices at the same time he was an army intendant. In the period before 1660 when many intendants were also masters of requests, they received income from this office. Precise salary figures are rare, but in the seventeenth century masters of requests averaged around 4,000 *livres* annually from their office, while one study gives a more accurate figure for the eighteenth century: 2,600 to 4,900 *livres* a year. Certainly not all army intendants held this office, but they frequently held others. Many possessed the position of councilor of state, which averaged 3,000 to 5,500 *livres* a year in the eighteenth century, depending upon the type of appointment (*semestre*

5. Vivian R. Gruder, *The Royal Provincial Intendants: A Governing Elite in Eighteenth-Century France* (Ithaca, N.Y., 1968), pp. 237–40 (hereafter cited as *Provincial Intendants*); "Ordonnance de 3000# pour M Balthazar allant faire la charge d'intendant en l'armée de Lombardie," A.G. A¹ 105, fol. 267; similarly for Argenson, A.G. A¹ 88, fol. 35; the king to Servien, 11 July 1647, A.G. A¹ 104, p. 23rv; Robert to Louvois, 29 May 1673, A.G. A¹ 338, fol. 186.

6. "Don pour le Sr de Choisy," A.G. A¹ 121, fol. 196; gift to Servien, A.G. A¹ 141, fol. 194; "brevet et don en faveur de Sr Talon," A.G. A¹ 152, fol. 405.

or *ordinaire*). If an army intendant held a joint appointment as intendant of an army and a province, his salary naturally increased. This doubling or even tripling of office was not unusual in the seventeenth century, and an army intendant might have an income of 30,000 *livres* a year.[7]

It is doubtful whether an army intendant could make his fortune from the income of his office, for there were large expenses in an intendancy. When Michel Le Peletier de Souzy left his provincial intendancy of Flanders in 1683, he had 11–12,000 *écus* of debts. The situation might be even worse for army intendants in the first half of the seventeenth century, for these men often lent money or arranged loans on their own credit for the support of the army. This was especially true during the government's financial difficulties before and during the Fronde. Sometimes the state reimbursed its intendants, sometimes not. In 1650 Ennemond Servien received 6,000 *livres* from the king for the cost of hospital supplies that he had paid for out of his own pocket. But not all army intendants were so lucky. Pierre de Pontac lent the crown more than 50,000 *livres* to pay for the army of Guyenne's expenses during the Fronde. Not only did the crown default on this, but it made no move to pay Pontac's back salary. Obviously only men with substantial fortunes could afford this type of service.[8]

This is not to say that an army intendancy might not be profitable, especially if an intendant was an opportunist. Jean-Baptiste Colbert once accused Louis Robert of having profited from a contract he had negotiated for the army of Italy. Robert was the same agent that Louvois used to squeeze every bit of money from Holland during the Dutch War. The intendant was efficient in exacting vast sums from the occupied area, and undoubtedly some of it stuck to his palms. It

7. Croquez, *Intendance de Flandre wallonne*, p. 49; Gruder, *Provincial Intendants*, pp. 237–40. Although I uncovered no figure for the total annual income of an army intendant, that of a provincial intendant will give some approximation. Michel Le Peletier de Souzy served as intendant of Flanders from 1668 to 1683. He made 12,000 *livres* a year as intendant, 4,000 *livres* as master of requests; he also received gifts from the crown and province varying from 20,000 *livres* in 1671 to 6,000 *livres* in 1681. After 1670, Souzy was *surintendant des Monts-de-piété*, which brought him an additional 12,000 *livres* annually. On the average, the intendant earned about 70,000 *livres* per year (Croquez, *Intendance de Flandre wallonne*, pp. 49–50).

8. Croquez, *Intendance de Flandre wallonne*, p. 50; rough draft of an order to pay Servien 6,000 *livres*, A.G. A¹ 122, fol. 429; Pontac to Le Tellier, 4 November 1652, from Agen, A.G. A¹ 134, fol. 354. Mousnier, *Lettres au Séguier*, 1:134–35, recounts the tale of Jacques de Chaulnes, who requested authorization to resign his office of master of requests in favor of his brother-in-law, because Chaulnes's income as intendant covered only half of his expenses.

is also doubtful whether a man of humble antecedents like Jacques Charuel could have acquired a fortune in excess of 350,000 *livres* without knowing how to benefit from his situation.[9]

Aside from their salaries and various gifts, most intendants had other forms of wealth: *rentes*, lands, and offices obtained through family inheritance or prosperous marriages. It is difficult to sketch more than a fragmentary picture of this type of wealth, but scattered receipts from the payment of *rentes* remain in the genealogical records of the Bibliothèque Nationale. They indicate that the intendants often invested money over long periods of time, for *rentes* were secure bonds that paid a fixed rate of interest. Generally they were drawn upon the Hôtel de Ville of Paris, the guarantor, which paid interest from the profits of state monopolies, such as the *gabelles*. Six receipts of Louis Robert, acknowledging payment from such *rentes*, exist today. The total amount of money constituted in these bonds is slightly more than three thousand *livres*, but they were established at six different dates: 1552, 1563, 1572, 1635, 1682, and 1699, and undoubtedly represent inherited wealth in the Robert family. It is hazardous, however, to make much of such documents, because no complete collection of *rente* records remains. About all that can be said is that investment in *rentes* was a common occurrence among the Parisian magistracy. Nevertheless, there is no indication that this type of investment reached the heights it did in the eighteenth century, when some magistrates had so many *rentes* they needed to employ secretaries to administer their portfolios.[10]

Land was another important source for investment in the ancien régime. Even for bourgeois fortunes, often more than half of their wealth was in the form of land ownership. In any agrarian society, and France was predominantly agricultural in the seventeenth century, land was important both as a source of wealth and for social prestige. It stressed the continuity of tradition and stood as a symbol for social preeminence, and many attempted to conceal a recent or rapid climb up the social ladder by the possession of extensive estates.[11]

Army intendants of old seignorial families owned a great deal of

9. Pierre-Adolphe Chéruel, ed., *Journal d'Olivier Lefèvre d'Ormesson et extraits des mémoires d'André Lefèvre d'Ormesson,* Collection de documents inédits sur l'histoire de France (Paris, 1860–61), 2:55. For Robert's later activities, see Camille Rousset, *Histoire de Louvois et de son administration politique et militaire* (Paris, 1862–63), 1:435–42. For Charuel's fortune, see n. 3, above.

10. Six *quittances des rentes* of the Hôtel de Ville, B.N. *pièces originales* 2499–2500, family "Robert," fols. 151 and 2, 3, 14, 15, 16, respectively; J. H. Shennan, *The Parlement of Paris* (Ithaca, N.Y., 1968), p. 146.

11. Cf. Pierre Goubert, *Cent Mille Provinciaux au XVIIe siècle: Beauvais et le Beauvaisis de 1600 à 1730* (Paris, 1968), pp. 377–78.

property. For example, René de Voyer d'Argenson possessed the seigniories of Argenson, La Bailliolière, and Chastres in Touraine worth 80,000 *livres*, the seigniory of Vueil-le-Mesnil in the *bailliage* of Blois of undetermined value, and the seigniories of La Loyre and de Moulay in Berry worth 50,000 *livres*, plus various houses in rue Vieille-du-Temple and one in rue des Pouliers in Paris. But even "new men" such as Jacques Charuel, whose father had been a receiver of the seigniory of Louvois and clerk (*greffier*) of the *bailliage* and provostship of Epernay, acquired extensive real estate. Charuel owned the seigniory of Varennes between Lagny and Meaux worth 40,000 *livres*, the meadows (*herbages*) of Grealle near Pont l'Evêque in Normandy worth 80,000 *livres*, the meadows of Vitres that he purchased from a monsieur de Laigle in 1671 for 12,500 *livres*, and the meadows of Aubreuil that he obtained in 1687. A similar situation existed for other army intendants. The legal documents of the period are full of references to estates; for example, Pierre Goury, intendant of the army in Catalonia in 1645, entitled himself in one contract "Pierre Goury, chevalier, viscount d'Enal, seignior of Mazurier, Bazoches, les Haultes, and other places. . . ." Even with such fragmentary evidence, it is reasonable to conclude that landholding was an important source of wealth for army intendants, as it was for other prominent officials, and a symbol of position in an aristocratic society.[12]

Officeholding constituted another form of investment in the ancien régime. With the payment of an annual tax, the *paulette*, most offices were the property of the officeholder and could be bought and sold like any other commodity. Usually Parisians made the sale within their families or a tight circle of friends, rarely to strangers. The closed market and the great inflation of prices for offices in the seventeenth century meant that families might tie up considerable sums in their offices, sums that might not be commensurate with an office's annual salary. In this situation offices represented a large amount of frozen assets that only the wealthy could afford. François Cazet de Vaux-torte, sometime army intendant in the 1640s, sold his office of advocate general in the *grand conseil* in 1639 for 111,000 *livres*, while Louis Robert held the office of president of the *chambre des comptes* in Paris, worth roughly 100,000 *écus* when he was forced to sell it in 1690 to pay off his gambling debts.[13]

12. Mousnier, *Lettres au Séguier*, 1:105; B.N. *dossiers bleus* 171, family "Charuel," fols. 2–13; contract between Gabriel de Rochechouart, duc de Monte-mart, and Pierre Goury, B.N. *pièces originales* 1378, family "Goury," fol. 30.

13. Mousnier, *Lettres au Séguier*, 1:159; Philippe de Courcillon, marquis de Dangeau, *Journal du marquis de Dangeau*, ed. Soulié et al. (Paris, 1854–60), 3:263, entry for 20 December 1690.

Even basic offices such as that of master of requests involved considerable amounts of capital. Before Louis Le Maistre de Bellejamme
served as army intendant, he purchased a mastership of requests from
Jean Amelot in 1626 for 159,000 *livres,* paying 4,000 *livres* in cash and
the balance in four other payments in 1626, 1629, 1631, and 1634. He
then turned around and sold the office for 158,000 *livres* in 1641. The
parents of future army intendant Alexandre de Sève gave their son
162,000 *livres* to purchase his office of master of requests upon the
occasion of his marriage in 1637.[14]

High parlementary offices were even more expensive. Jean-Edouard
Molé, sieur de Champlastreux, became a master of requests in 1643,
served as intendant of the armies of the prince of Condé in the 1640s, first
in Champagne, then in Germany, before becoming president *à mortier*
of the Parlement of Paris in 1653. Unfortunately there are few figures
available on the inflation of Parisian parlementary offices, but a study
of the same office in the Parlement of Rouen showed the following
increase in price: 40,000 *livres* in 1600; 100,000 in 1609; 120,000 in
1623; and 186,000 in 1632. It is reasonable to assume that the price of
the office in the Parisian court was much higher, for the Parlement of
Paris was the most prestigious of all the sovereign courts. Of course
Molé could afford it, for his father was a first president in the Parlement of Paris and keeper of the seals. Nevertheless, these prices illustrate the wealth in circulation at this level of society.[15]

One source of this wealth was marriage, for a carefully chosen match
often brought a handsome dowry into the family. An examination of
the occupations of sixty-five fathers of wives of army intendants reveals that the majority occupied important government positions and
had considerable wealth.[16] Eleven wives were daughters of *secrétaires du*

14. Mousnier, *Lettres au Séguier,* 1:153–54, 127.

15. Roland Mousnier, *La Vénalité des offices sous Henri IV et Louis XIII,*
2d ed., rev. and enl. (Paris, 1971), p. 363.

16. This and all the following genealogical details in this chapter are taken from
these sources: the manuscript records in the so-called *cabinet des titres* collection
of the Bibliothèque Nationale, including the *dossiers bleus, nouveau d'Hozier,
pièces originales, cabinet d'Hozier,* and *carrés d'Hozier; Biographies des maîtres
des requêtes depuis 1575 jusqu'en 1722, suite des généalogies des maîtres des
requêtes de François Blanchard,* manuscript in B.N. *fonds français* 14,018; plus
these published sources: Prevost and Roman d'Amat, *Dictionnaire de biographie
française;* Mousnier, *Lettres au Séguier,* 1:65–168 (individual notices), and 2:1183–
1227; François Bluche, *L'Origine des magistrats du Parlement de Paris au XVIII*
siècle, Mémoires de la Fédération des sociétés historiques et archéologiques de
Paris et de l'Ile-de-France, tomes 5–6 (Paris, 1956); Henri de Frondeville, *Les
Présidents du Parlement de Normandie (1499–1790)* (Paris and Rouen, 1953) and
Les Conseillers du Parlement de Normandie au seizième siècle (1499–1594) (Paris
and Rouen, 1960); Jean-Paul Charmeil, *Les Trésoriers de France à l'époque de la
Fronde: Contribution à l'histoire de l'administration financière sous l'ancien ré-*

roi, a largely honorific office that carried little responsibility but that was eagerly sought by a nouveau riche in the seventeenth century, for it carried much prestige and the advantage of immediate ennoblement.[17] The post was a sure indication of a family's fortune.[18] Thirteen intendants married daughters of high fiscal officials and financiers, including eight daughters of important treasurers, positions usually accompanied by large personal fortunes.[19] Another nineteen wives were children of members of the sovereign courts, the highest French judiciary officials, and ten of these nineteen were daughters of men in the two sovereign courts that judged financial matters, the *chambre des comptes* and the *cour des aides*.[20] One intendant married a daugh-

gime (Paris, 1964) (hereafter cited as *Trésoriers de France*). I decided not to footnote specific page numbers, for this would make the chapter too bulky; in this I follow the practice of Mousnier, *Lettres au Séguier*. It is difficult to pinpoint specific offices, because men usually held several during their careers. I have taken the most important office, usually held in the latter years of a man's life, but since not all the genealogical records are complete, even with cross-checking the possibility for error is great.

17. The wives of Philibert Baussan, Claude Bazin de Bezons, Jacques Bigot, Louis I Chauvelin, Honoré Courtin, Pierre Goury, Denis de Heere, Claude de La Fond, Louis Machaut, Louis Robert, and René de Voyer d'Argenson were daughters of *secrétaires du roi*. Charles Machaut's second wife was not a daughter, but a widow of a secretary before her remarriage.

18. For the importance of the office of *secrétaire du roi*, see Gruder's brief comment, *Provincial Intendants*, p. 110, and the more lengthy article in Marion, *Dictionnaire des institutions de la France*, s.v. "secrétaire du roi."

19. Eight intendants married daughters of important treasurers, including four *trésoriers de l'épargne:* Guillaume Bautru, count of Serran, Claude Gobelin by his first marriage, Jean-Jacques de Mesmes, and Gédéon Tallemant. Another, Jean-Edouard Molé de Champlastreaux, married the daughter of a *trésorier des parties casuelles;* one, François Bazin, the daughter of a *trésorier de l'extrordinaire de la guerre;* and two, Jacques Amelot de Beaulieu and Charles Machaut, the daughters of *trésoriers-généraux de France*. Additionally, Gaspard Du Gué's second wife was the widow of a treasurer in the bureau of finance at Lyon (Amelot de Beaulieu's father-in-law was also a *secrétaire du roi* as well as a treasurer).

Five other intendants married daughters of fiscal officials and financiers: Antoine Bordeaux married the daughter of a receiver of finances at Tours, Charles Colbert de Croissy married the child of a financier and her marriage portion was a million and a half *livres*, Jean Colbert du Terron married the child of a receiver of the *gabelles* in Champagne, Denis Marin's father-in-law was a *partisan* or tax farmer, while François Villemontée married the offspring of a paymaster of the *rentes* of the Hôtel de Ville (Louis I Chauvelin's father-in-law also served as an important tax farmer—a *fermier général*—before obtaining the office of *secrétaire du roi*).

20. Nine army intendants married daughters of officials of the Parisian *chambre des comptes:* Simon Arnauld de Pomponne, Gilbert Colbert de Saint Pouenges, Antoine Le Fèvre de la Barre, Louis Le Maistre de Bellejamme in his first marriage, Jacques Paget, and Auguste-Robert de Pomereu all married daughters of ordinary *maîtres des comptes*, while Olivier Le Fèvre d'Ormesson and Geoffrey Luillier d'Orgeval married daughters of presidents of the *chambre des comptes* in Paris. François Bochard de Sarron Champigny's father-in-law was a *procureur général* in the same court. (Bluche, *Origine des magistrats*, p. 267, claims that

ter of the possessor of an honorific position at the royal court.[21] Ten
married daughters of the highest government officials: councilors of
state, masters of requests, and intendants.[22] Ten wives were offspring
of lesser judicial, administrative, and military officials, mostly provincial.[23] Only one intendant married the daughter of a non–government
official, a doctor of medicine, belonging to a profession usually pos-
sessing some wealth.[24]

When one examines the wealth that these women brought into marriage, it is obvious why intendants, like other officeholders, preferred
daughters of financial officials and rich men. René de Voyer d'Argenson inherited a total of 150,000 *livres* from his wife and her family,
while the two wives of Louis Le Maistre de Bellejamme each provided
70,000 *livres* as dowries. Guillaume Bautru's wife was not very pretty,
but she brought 400,000 *livres* to her marriage bed—surely a compen-

Louis Le Maistre de Bellejamme's first wife was the daughter of a *maître des
requêtes* and not a *maître des comptes*, but see Jean-Baptiste de l'Hermite-Soliers
and François Blanchard, *Les Eloges de tous les premiers présidens du Parlement
de Paris, depuis qu'il a esté rendu sédentaire jusques à présent. Ensemble leurs
généalogies, épitaphes, arms et blazons, en taille douce* [Paris, 1645], pp. 337–45,
which states that the father was a *maître des comptes.*)

Michel Le Tellier, secretary of state for war and former army intendant, married a daughter of an official in the *cours des aides*. The other nine intendants
who married daughters of members of the sovereign courts are the following:
Jean Balthazar and Jacques Dyel de Miromesnil had fathers-in-law who were
advocates in the *grand conseil*. Daniel Voisin's first wife and that of Nicolas
Fouquet were daughters of presidents of the chambers of the Parlement of Paris
known as the *requêtes du palais*, while Voisin's second wife was the daughter of
Omer Talon, advocate general of the Parlement of Paris. Jean Lauzon, sieur de
Liré, and Michel Le Peletier de Souzy married offspring of councilors of the
Parisian parlement. Gaspard Du Gué's second wife was the daughter of a president
à *mortier* of the Parlement of Grenoble, and Pierre de Pontac married the daugh-
ter of a *procureur général* of the Parlement of Toulouse.

21. Nicolas Bretel de Grémonville married the daughter of a *secrétaire d'état
de la couronne de Navarre et cabinet du roi.*

22. Dreux d'Aubray, Paul Barillon d'Amoncourt, Jean-Etienne Bouchu, Fran-
çois Cazet de Vauxtorte, Jacques de Chaulnes, Jean de Choisy, Dreux-Louis Du
Gué de Bagnols, Louis Laisné sieur de la Marguerie, Louis-François Le Fèvre de
Caumartin in his second marriage, and Charles Mouceau de Nollant.

23. Pierre Imbert's wife was the daughter of a captain in the regiment of Du
Plessis Praslain; René Joüienne d'Esgrigny's wife was the daughter of a *commissaire général* of artillery; Isaac Laffemas's two wives were both offspring of
notaries, one in the Châtelet; François Lasnier married the child of a *procureur
du roi* in the provosty of Angers who was also an *échevin*, or city councilor, in
the town of the same name; Louis-François Le Fèvre de Caumartin's first wife
was the daughter of a lieutenant general of the *présidial* of Poitiers; Charles Le
Roy de La Potherie married the daughter of a *maître des eaux et forêts* in Maine;
François Le Tonnelier de Breteuil's father-in-law was a *lieutenant du roi* in Artois
and commandant of Hesdin; René Le Vayer's spouse was the child of a *lieutenant
criminel* in Le Mans; and Ennemond Servien married the offspring of a *bailli* of
Valence in Dauphiné.

24. Michel-Louis Malézieu's father-in-law was a physician.

sation for lack of pulchritude. The dowry of Jacques de Chaulnes's wife was 75,000 *livres tournois,* while that of Denis de Heere's spouse brought 60,000 *livres.* Charles Le Roy de La Potherie's wife contributed 60,000 *livres* by her marriage contract, half the sum of François Cazet de Vauxtorte's bride. The daughter of a member of the *chambre des comptes* in Paris offered 200,000 *livres* as part of her marriage settlement with Antoine Le Fèvre de la Barre. Charles Colbert de Croissy's spouse brought 1,500,000 *livres* as a dowry, surely a record price even for a brother of the great Colbert.[25]

Table 1
Occupations of Intendants' Fathers-in-Law

1	position at court
11	*secrétaires du roi*
10	high government officials (councilors of state, masters of requests, intendants)
19	members of sovereign courts, including:
	1 *cour des aides*
	9 *chambre des comptes*
	9 parlementarians
13	treasurers and important fiscal officers
10	lesser judicial, administrative, and military positions
1	medical doctor
65	Total

Officeholding and the expense of public service required substantial sums in the seventeenth century, and few could afford to overlook the *dot* of a prospective bride. Army intendants, in particular, needed a large fortune in the first half of the century, because the government expected the intendants to lend money or use credit to supply the army in times of emergency. This was particularly true during the troubled period known as the Fronde. But by the second half of the century, the crown was on a much better financial footing—no longer did intendants need to advance their own fortunes in support of the army. This fact coincided with another circumstance: fewer intendants were drawn from the ranks of older families with large fortunes and a tradition of governmental service. Increasingly the army intendants were drawn from the ranks of "new men," former *commissaires des guerres* trained in military service, from solid bourgeois and lesser administrative backgrounds, without the large fortunes of established families such as the Arnaulds, the Argensons, the Aligres, the Molés,

25. Mousnier, *Lettres au Séguier,* 1:105, 153, 134, 150, 159; Prevost and Roman d'Amat, *Dictionnaire de biographie française,* s.v. "Bautru (Guillaume)" and "Colbert, Charles (marquis de Croissy)." B.N. *dossiers bleus* 267, family "Fèvre (Le)."

or the Le Fèvres. By the latter half of the century, such men as Carlier, Charuel, Camus de Beaulieu, Mouceau de Nollant, Malézieu, and Heisse had made their fortunes in the army, not in regular officeholding. They also tried to erase the marks of their base origin and join the ranks of the aristocracy. This vertical mobility was not unusual; it was part of a continuing process in the ancien régime. Generally it took four or more generations for a family to complete the climb, but the time might be shortened by luck or ability.[26]

The origin of almost all the army intendants' families was bourgeois; only a few were of noble descent. Jean-Jacques de Mesmes, for example, traced his family back to a knightly ancestor living in 1219, but this nobility is certain only from the fifteenth century. The Voyer d'Argenson family was ennobled in 1375 and exercised the profession of arms until the seventeenth century, when it entered the "robe." Jacques de Chaulnes's brother produced proof of nobility going back to a knight living in 1270, but a recent historian has questioned this claim.[27] All the other family origins were nonnoble. Unfortunately only a few first ancestors of these families are known today. Two were lawyers (Le Maistre de Bellejamme and Bouchu), two were notaries (Cazet de Vauxtorte and Arnauld), two were clerks (Aligre and La Potherie), and fourteen were merchants: Amelot (herrings), Aubray, Bretel de Grémonville, Charuel, Choisy, Colbert (cloth), Fouquet (cloth), Heere (silk), Le Gendre, Le Peletier de Souzy (furrier), Le Tellier (cloth), Molé (cloth), Pomereu, and Sève. One first ancestor was a skilled artisan, a dyer (Gobelin), another a gendarme in the royal guards (Joüenne d'Esgrigny), a third an apothecary (Barillon), but only one descended from the son of a *laboureur*, a rich peasant, who had escaped from the farm to become a doctor (Mouceau de Nollant).

The provincial origin of almost all these families is certain. Perhaps only six were native to Paris. The Courtins were living in the capital in the fourteenth century, the Bazins and the Gobelins in the fifteenth, while the Le Telliers, Bezonses, and Baussans date from the sixteenth century. By the seventeenth century most of the other families had migrated to Paris, the font of all honors and advancement, and identified themselves with the Parisian way of life.[28]

---

26. Concerning the "new men," cf. Bluche, *Magistrats du Parlement*, p. 96, who remarks that although the expression is sanctified by common usage, it is only partially satisfying, for most of the new men arose from the stratum of the lesser judicial and administrative officials well below the level of the older families; but they were not from the lowest ranks of society as might be expected by the term.

27. Mousnier, *Lettres au Séguier*, 1:129.

28. Cf. the remarks in ibid., 1:55–56. The La Potherie family came from Normandy, while the Le Peletiers originally lived in Le Mans. The Le Fèvre

Fifteen army intendants, however, either were not native to the capital or were newly Parisian. Jacques Charuel was born in Epernay in 1616, where his father was a minor official. Gaspard Du Gué, sieur de Bagnols, was a treasurer of France at Lyon, although his family had lived in Moulins for several generations. Pierre Gargan was a native of Châlons-sur-Marne before becoming *secrétaire du roi*, intendant of finance, and serving briefly as army intendant. Denis Marin, another intendant of finance who acted as army intendant during the Fronde, was an apprenticed cobbler at Auxonne, but he did not have enough money to become a master so he went to Paris and worked as a valet of the prince of Conty, before working his way up the ladder to success. Pierre de Pontac, first president of the *cour des aides* of Bordeaux, served at the same time as intendant of Harcourt's army in Guyenne. Louis Gombault's family lived in Troyes, where he later served as councilor in the *bailliage* and *siège présidial*, before serving as army intendant in Catalonia. Nicolas Bretel de Grémonville was born in Rouen, where four generations of his forefathers had been presidents of the Parlement of Normandy. Jacques Dyel de Miromesnil was another native of Rouen. The Servien family were inhabitants of Grenoble, and Ennemond Servien, brother of the minister Abel Servien, held various offices in Dauphiné before serving as intendant of the army of Italy and ambassador to Savoy. Robert Trobat, army intendant in Catalonia, was a native of nearby Perpignan. René Le Vayer's family lived for several generations in Le Mans, where the future army intendant held the position of lieutenant general of that *sénéchausée*. Lack of documentation prevents knowing when François Lasnier came to Paris, but his father was a mayor of Angers and president in the local *siège présidial*. François Cazet de Vauxtorte became a Parisian sometime before his marriage in 1634. Etienne Carlier's father and grandfather had both lived and married in the town of Clermont, although

de Caumartins lived in Ponthieu, the Molés in Troyes, and the Colberts in Rheims. The Aligres descended from bourgeoisie in Chartres, and the Pomereu family originated in the Soissonnais. The Talons claimed descent from an Irish army officer, but were really modest bourgeoisie in Picardy during the sixteenth century. The Heere family came from Bourges. The Bautru family lived in the province of Anjou, the Arnaulds and Villemontées in Auvergne. The Bouchus and the Bocharts de Champigny were natives of Burgundy. The Mouceau family originated in Châteaudun. The Barillons were a bourgeois family of Issoire in the fifteenth century, while the Le Maistres de Bellejamme were a bourgeois family in Montlhéry that same century, but the Amelots were still bourgeoisie in Orléans in the sixteenth century. The Robert family claimed descent from the Roberts of Orléans. The Voisin family was native to Tours, the Chauvelins to the Vendômois, and the Machauts to the Rethélois. The Le Fèvre d'Ormesson family lived in Ormesson near Montmorency before becoming bourgeois *robins* of Paris at the end of the fifteenth century. The Aubrays were merchants of Conflans-Sainte Honorine in the following century.

he and his brother, Pierre, were Parisians. Guillaume de Sève moved to Paris, and his son Alexandre, the army intendant, was born in the capital, but he can hardly be considered a Parisian of long standing.[29]

Special circumstances explain a number of these non-Parisians. Gargan, Marin, and Pontac were fiscal officials used as intendants during the period of the Fronde. Others such as Gombault in Troyes, Servien in Dauphiné, and Trobat in Roussillon were local magistrates of some importance, intimately acquainted with conditions on the frontier. In this respect, they were logical candidates to serve as army intendants. These men remained a minority, however; the overwhelming majority of army intendants were Parisian albeit of provincial ancestry.

In a society in which the aristocracy held the highest privileges, ennoblement was an important step upward—indeed, a prerequisite for admission to the top positions. Certainly every one of the army intendants already possessed or else obtained the legal status of nobleman during his lifetime, but there was wide variation in the methods of ennoblement.[30] First, there were *familles chevaleresques*, descended from ancient knights, who had borne arms in the king's service and possessed nobility by birth. This group is sometimes called the "nobility of the sword," but the term "chivalric nobility" is probably more appropriate. Only two or three army intendants belonged to this group: Argenson, Mesmes, and perhaps Chaulnes, although even then the Voyer d'Argenson family had been ennobled by letters patent in 1375, but were considered *chevaleresque*. Next, there were "families of ancient extraction" that could not prove their nobility by evidence predating the sixteenth century, but nevertheless were accepted as noble by the common consensus of their provinces. This sometimes led to difficulties because the state made frequent searches for false nobles, particularly as people claimed to be noble in order to escape the tax rolls. To prevent this occurrence, some families submitted their claims to fiscal courts for certification. Thus the Parisian *cour des aides* upheld the nobility of the Bezons family in 1612.

Beneath this group were the created noblemen. The most honorable method of ennoblement was by letters patent from the king supposedly in reward for services to the crown, although some were purchased (unfortunately it is difficult to determine who bought their titles; but

29. Julien Pietre, former canon at Amiens, then treasurer of France in that city, might be considered a non-Parisian, but his grandfather was a doctor of medicine in Paris.

30. There are, of course, many different classifications of the French aristocracy. See, for example, the remarks of Bluche, *Magistrats du Parlement*, pp. 87–89. I have essentially followed Bluche's division in the following paragraphs. Cf. Gruder, *Provincial Intendants*, pp. 117–27.

a member of the La Potherie family achieved nobility in this way for 550 *livres* in 1542). Next came ennoblement by officeholding: by the position of *secrétaire du roi*, by important judicial and fiscal offices (the so-called nobility of the robe), or by various municipal offices (the so-called nobility *de cloche*). Finally, there was incorporation into the nobility, *familles agrégées à noblesse*, by the possession of fiefs and living nobly until one was recognized as such. Technically this was a semiusurpation, because the mere possession of a fief did not necessarily confer nobility.

Officeholding remained the most common method of ennoblement. Some offices conferred the immediate legal status of nobility, either for the possessor alone or, in certain cases, for his entire family. Other offices granted only partial nobility, since they had to be held for several generations before a family could call itself noble. Because most offices were venal, an ambitious bourgeois might purchase the appropriate office and join the ranks of the second estate. At least eight families of army intendants obtained nobility through the acquisition of the office of *secrétaire du roi*: Barillon in 1534, Le Maistre de Bellejamme in 1562, Pomereu in 1586, Choisy in 1592, Marin in 1632, Le Vayer in 1636, Le Peletier in 1637, and La Fond in 1650. Membership in the Parlement of Paris or any of the other sovereign courts also conferred nobility. Parlementary positions ennobled the Bocharts de Champigny in 1466, the Molés in 1537, the Cazets de Vauxtorte in 1585, and the Bretels de Grémonville in 1588, while membership in the *chambre des comptes* at Paris granted the same honor to the Le Telliers in 1574. The office of treasurer of France ennobled Jean Le Fèvre de Caumartin and his family in 1555. There is also an example of the nobility *de cloche* among the families of army intendants, for Jean de Sève gained the ranks of the aristocracy through the office of *échevin* or town councilor of Lyon in 1510. Many other families were ennobled by officeholding, but the particulars are not known: the Machauts in 1523, Mouceaus in 1556, Le Fèvres d'Ormesson in 1568, Amelots in 1580, Aubrays in 1582, Chauvelins in 1634, and Baussans and Le Gendres in the middle of the seventeenth century.[31]

Some army intendants, particularly those in the latter half of the seventeenth century, had difficulty obtaining recognition of their nobility. Louis Robert and his two brothers submitted an elaborate genealogy going back to one Antoine Robert, *secrétaire du roi* in the days of Louis XI, and his descendants, noble doctors of the University

31. Bluche's genealogical dictionary, *Origine des magistrats du Parlement de Paris*, is particularly important for determining the dates of ennoblement by officeholding.

of Orléans. When this proof was presented to the commissioners for the search of usurpations of the title of nobility, they rejected the claim, contending that practically every document submitted as evidence of nobility was either false or irrelevant. Nonetheless, Robert's claim was accepted by a decree from the royal council, dated 4 May 1667, which acknowledged the Roberts as noble. Undoubtedly Robert's service as army intendant helped his case, and he had friends in high places—Le Tellier was his protector.[32]

Another of Le Tellier's clients, Etienne Carlier, had to go to the crown for a letter of confirmation substantiating his nobility. This document accepted the specious claim that the Carliers descended from a noble family in the Low Countries that had fled their native land when Don Juan of Austria suspected them of spying for the French. Because of the family's flight, it could not justify its nobility, for the enemy had confiscated and burned the necessary documents, and economic conditions had forced the family to live bourgeois for several generations. Although the crown accepted Carlier's story, an anonymous genealogist disagreed, and his opinion is recorded in the manuscript collection known as the *dossiers bleus*. He claimed that there was no evidence tracing the Carliers back to the Netherlands, and he accused them of deliberate falsification. Nevertheless, they obtained their letter of confirmation in 1668 and were again upheld in their nobility by a decision in 1670. Once more the power of the secretary of state for war had worked wonders.[33]

Not all army intendants attempted to conceal their recent origin. For example, the Heisse family records in the genealogical collection of the Bibliothèque Nationale contain a copy of an ordonnance of 1697 registering the arms of Nicolle Brossier, widow of Jean Heisse and mother of Thomas, an army intendant, in the Armorial of France. These arms originally had been given to Jean by Emperor Leopold. Among the fragmentary records of the Malézieu family are naturalization papers for Maria-Jéronime de Marcarty, wife of army intendant Michel-Louis Malézieu. One of the d'Hozier family, keepers of the armorial records, wrote on this document: "the arms which Charles d'Hozier, my granduncle, gave to this Maria-Jéronime de Marcarty . . . are on a field of sinople, a golden eagle having on its breast a

32. B.N. *nouveau d'Hozier* 286, family "Robert," fols. 29–35.
33. B.N. *cabinet d'Hozier* 78, family "Carlier," fol. 2, for the confirmation of the Carlier nobility, but see in B.N. *dossiers bleus* 154, family "Carlier," fol. 6, the comment of the anonymous genealogist: "Voila le fondement de la noblesse de cette famille, car audessus desd. lettres de 1668 il y a des notes que la plus grande partie des titres de cette famille sont faux." Indeed, the Carlier family records in the B.N. are so confused that it is difficult to decipher them.

crescent of scarlet charged with a spur of silver." Another document noted that this same granduncle granted Malézieu the following arms: on a field of azure, a golden chevron accompanied in the chief by two golden lilies with a golden lion posed under the chevron. The army intendant Jacques-Louis Le Marié obtained his own arms during his lifetime too, three black hands on a field of gold. All three men, Heisse, Malézieu, and Le Marié, were former *commissaires des guerres* who rose to the rank of army intendant. They were not members of the old officeholding families; they were "new men" climbing into the ranks of nobility by their service to the crown.[34]

The change in occupation of the fathers and grandfathers of the army intendants illustrates this process of upward mobility, whether already under way for several generations or just begun by "new men." The professions of fifty-eight grandfathers are known. Four lived nobly as seigniors, including two who held positions at court as gentlemen of the king's bedchamber.[35] Three grandfathers had important government positions as masters of requests and members of the *conseil du roi*. One of the three even became a keeper of the seals.[36] Six grandfathers enriched themselves and acquired the honorific position of *secrétaire du roi*.[37] Twenty-five others functioned in the sovereign courts and included one official in the Parisian *cour des aides*, three in the *grand conseil*, four in the *chambre des comptes* in Paris, and seventeen in the parlements.[38] Another seven held important

34. B.N. *pièces originales* 1502, family "Heiss," fols. 4–5; B.N. *carrés d'Hozier* 404, family "Malézieu," fols. 272–73; B.N. *pièces originales* 1853, family "Le Marié," fols. 29–30.

35. Jacques Goury, Pierre's grandfather; Jacques Le Roy de La Potherie, Charles's grandfather; Jean de Voyer, viscount of Paulmy and grandfather of René de Voyer d'Argenson; and René de Voyer, viscount of Paulmy and grandfather of René de Voyer, sieur de Dorée. The two Voyers, who were brothers, held positions as gentlemen of the king's bedchamber.

36. Jean Bochart served as master of requests and member of the royal council; he was the ancestor of François Bochart, sieur de Sarron Champigny. Jean-Jacques de Mesmes's grandfather, also named Jean-Jacques, was a master of requests. Louis-François Le Fèvre de Caumartin's grandfather Louis, a former intendant and ambassador, ended his career as keeper of the seals.

37. The six *secrétaires du roi* are: Claude d'Aubray, grandfather of Dreux; Oudard Colbert, former merchant at Troyes and grandfather of Gilbert Colbert de Saint Pouenges; Antoine Le Fèvre, clerk in the bureau of finances in Paris and grandfather of Antoine Le Fèvre de la Barre; Pierre Le Maistre, who served as chief clerk in the *chambre des comptes* in Paris before becoming secretary, the ancestor of Louis Le Maistre de Bellejamme; Jacques Pomereu, grandfather of Auguste-Robert; and Thomas de Pontac, former head clerk in the Parlement in Bordeaux and ancestor of Pierre.

38. Claude Le Tonnelier de Breteuil, grandfather of François and father of Louis, was a *procureur général* in the sovereign court known as the *cour des aides* in Paris. Guillaume Bautru, grandfather of Guillaume, count of Serran, was a councilor in the *grand conseil*, where Simon Camus, ancestor of the Camus broth-

fiscal offices, mostly as treasurers or tax collectors.[39] Three functioned in petty judicial and administrative positions, although there were probably others, for the genealogical records of the "new men" are

ers, served as *procureur,* and Nicolas Bigot, grandfather of Jacques, served as *procureur général.*

The four grandfathers who served in the *chambre des comptes* are these: Guillaume Boucherat, ancestor of Louis, served as an auditor; Michel Le Tellier, grandfather of the secretary of war, and Louis Machaut, grandfather of Louis and father of Charles (Charles was Louis's uncle), served as ordinary councilors, or *maîtres;* Olivier Le Fèvre, grandfather of Olivier Le Fèvre d'Ormesson, was president of the court.

The seventeen parlementarians are: Edouard Molé, grandfather of Jean-Edouard Molé de Champlastreux, president *à mortier* of the Parlement of Paris; Jean Amelot, grandfather of Jacques Amelot de Beaulieu and president of one of the *chambres des enquêtes;* Jean Courtin, Honoré's grandfather, a *doyen* through seniority. Two royal representatives served in the Parlement of Paris: Antoine Arnauld, grandfather of Robert Arnauld d'Andilly, *procureur général aux requêtes,* and Jean Baussan, Philibert's ancestor, also a *procureur.* Four grandfathers served as ordinary members, or *conseillers,* of the Parisian parlement: Jean Barillon, grandfather of Paul Barillon d'Amoncourt; Denis de Heere, grandfather of another Denis; Baptiste Machaut, grandfather of Charles; and Nicolas Fouquet's grandfather. Four ancestors worked as *avocats* in the Parlement of Paris: Simon Arnauld de Pomponne's grandfather Antoine, Michel Le Peletier de Souzy's grandfather Jean, Charles Mouceau de Nollant's grandfather Michel, and Louis Robert's grandfather Anne. Four grandfathers served in parlements outside Paris: Louis Bretel, forefather of Nicholas Bretel de Grémonville, was president *à mortier* of the Parlement of Rouen; Jean Bouchu, grandfather of Jean-Etienne, was first president of the Parlement of Dijon; Jean, ancestor of François Cazet de Vauxtorte, served as a councilor in the Parlement in Brittany; while Gérard, grandfather of Ennemond Servien, held the same position in the Parlement of Grenoble.

39. Guillaume Bordeaux, grandfather of Antoine and father of Guillaume, was a paymaster for the *chambre des comptes* in Paris. Gaspard Du Gué, treasurer of France at Lyon and sometime army intendant in his own right, was the grandfather of Dreux-Louis Du Gué de Bagnols. Louis Laisné de la Marguerie's grandfather was Cibard Laisné, receiver of *tailles* in Angoumois. Antoine de Chaulnes, treasurer of the extraordinary funds for war and later controller general, was the ancestor of Jacques de Chaulnes. Louis Chauvelin, father of Louis I and grandfather of Louis II, was a receiver of domains and forests in the Ile-de-France. Mathurin Sublet, grandfather of François Sublet de Noyers, was a receiver of *tailles* in the *élection* of Gisors in Normandy. Jean Colbert, former cloth merchant at Rheims, became controller general of the *gabelles* in the provinces of Picardy and Burgundy. He was the grandfather of Charles Colbert de Croissy and probably of Jean Colbert du Terron.

The genealogical records of the Colbert family are confusing, partly due to deliberate falsification; see Roland Mousnier, ed., *Le Conseil du roi de Louis XII à la Révolution,* Publication de la Faculté des lettres et sciences humaines de Paris-Sorbonne, series "Recherches," tome 56 (Paris, 1970), pp. 165–66. Some genealogical records in B.N. *dossiers bleus* 203, family "Colbert," fols. 1–27, indicate that Jean (sometimes the name Charles is given), Colbert du Terron's grandfather, was a lieutenant of the town of Rheims, but Prevost and Roman d'Amat's *Dictionnaire de biographie française,* s.v. "Colbert (family)" and "Jean (sieur de Terron, marquis de Bourbonne)," says that his father was the controller general of *gabelles,* and his grandfather was Oudard I, merchant at Rheims. Because I was unsure of his exact genealogy, I excluded him from this part of the study.

largely incomplete.[40] Three held municipal offices as mayors or city councilmen.[41] Finally, seven others are known to have been bourgeoisie.[42]

Table 2
Occupations of Intendants' Grandfathers

---

4 lived nobly (including court positions)
6 *secrétaires du roi*
3 high government officials (councilors of state, masters of requests, intendants)
25 members of sovereign courts, including:
  3 *grand conseil*
  1 *cour des aides*
  4 *chambre des comptes*
  17 parlementarians
7 treasurers and important fiscal officers
3 lesser judicial and administrative positions
3 municipal officers
7 bourgeoisie

---

58 Total

The overwhelming occupational fact of these ancestors is the number of high judicial figures, members of the sovereign courts, and par-

40. Nicolas Dyel was an *avocat du roi* in the viscounty of Arques in Normandy, then a lieutenant in the *bailli* of Caux, as well as grandfather of Jacques Dyel de Miromesnil. René Joüenne was a royal councilor in the *présidial* of Le Mans before siring the father of René Joüenne d'Esgrigny. François Villemontée's grandfather was a *procureur du roi* in the tribunal known as the Châtelet.

41. Jean-Baptiste Bazin, grandfather of François, was a city councilman, or *échevin*, of Paris, while Pierre de Sève, grandfather of Alexandre, was a city councilor in Lyon. Finally, Louis Gombault's grandfather Jean was a mayor in Troyes.

42. Claude Bazin de Bezons, grandfather of Claude, was a doctor regent of the faculty of medicine in Paris, while Simon Pietre, grandfather of Julien, was a dean in that same faculty. Jacques Charuel's grandfather was an apothecary in Epernay. Jean Bordeaux, merchant draper of Paris, elected consul of the merchants, was grandfather to Guillaume and great-grandfather of Antoine. The unnamed grandfather of Jean de Choisy was a wine merchant who followed the royal court. Etienne Carlier's ancestor, Louis, and Robert Talon, grandfather of the Talon family, were also bourgeois. The army intendants in the Talon family are also hard to pinpoint. According to the genealogical records in the Bibliothèque Nationale, *collection de Chérin* 192, family "Talon," fols. 2–6, the army intendants who served in the 1650s and 1660s were the children of Philippe, *avocat* and *bailli* of Châlons, and grandchildren of Robert, a bourgeois of the town. But the manuscripts in the war archives do not give any first names. Likely candidates, but impossible to prove, could be any or all of these brothers: Philippe, *commissaire des guerres;* Paul, *commissaire des guerres*, also identified as "intendant of Canada"; Jean, variously described as "intendant of Hainaut," "intendant of Quesnoy," and "intendant of Canada"; and Claude, intendant of Audenarde. Philippe Talon, from another branch of the family, is mentioned as an *intendant des vivres* in the army in Roussillon in 1677 and 1680, but I found no indication of him in my list of intendants.

ticularly parlementarians (twenty-five out of fifty-eight). They far outnumbered any other vocation and stand in sharp contrast to the few who occupied important positions in the royal administration as masters of requests. Almost all the grandfathers, however, held some type of judicial or administrative position. Discounting two doctors among the bourgeoisie, only six men engaged in commerce. Thus many families were firmly planted in the "office aristocracy," known as the *noblesse de robe*, two generations before their descendants served as army intendants.

Much more information is available on the occupations of the army intendants' fathers. Of the seventy-eight fathers on whom information is available, five lived nobly as aristocrats at court.[43] The number of male parents—compared to the number of grandfathers—who served in the highest administrative positions in the government increased to fifteen.[44] Seven acquired the office of *secrétaire du roi*,[45] while the

43. The seigniors and court officials were Geoffrey Luillier d'Orgeval, father of Geoffrey; Pierre de Voyer de Paulmy, father of René de Voyer d'Argenson; and Louis de Voyer, viscount of Paulmy, father of Dorée; all three served as gentlemen of the king's bedchamber. Two others, Jean Aligre, father of Michel Aligre de Saint-Lié, and Barthélemy Laffemas, father of Isaac, were valets of the king's bedchamber. Only nobles were admitted to the position of *valet de chambre* by the eighteenth century, but the situation was much freer in the early days of the seventeenth, particularly during the reign of Henry IV. Thus Barthélemy Laffemas began as a tailor, rose as businessman, and became a *valet de chambre* under Henry IV, before heading the king's commission to study commerce.

44. At least eleven fathers served as masters of requests and *conseillers du roi*. They included: Claude Bouchu, father of Jean-Etienne and long-time intendant of Burgundy; Jacques de Chaulnes, father of Jacques and dean of the masters of requests; Louis I Chauvelin, army intendant and father of Louis II; Jean-Baptiste Colbert de Saint Pouenges, father of Gilbert and intendant of Lorraine; Achille Courtin, father of Honoré; François Fouquet, father of Nicholas and ambassador to the Swiss; Guillaume Du Gué, father of Dreux-Louis de Bagnols; Jérôme Le Maistre, father of Louis de Bellejamme; Louis Le Fèvre de Caumartin, father of Louis-François, intendant of Picardy and ambassador to Venice; André Le Fèvre d'Ormesson, father of Olivier and intendant in Lyon; Louis Le Tonnelier de Breteuil, father of François and military and provincial intendant in his own name before becoming a controller general of finance.

Four other fathers do not seem to have been masters of requests: Robert Arnauld d'Andilly, father of Simon de Pomponne and onetime army intendant; Guillaume Bautru, father of Guillaume and ambassador to Spain, England, and Savoy; Guillaume Bordeaux, father of Antoine, former army intendant and intendant of finance in the government; and Antoine Le Fèvre de la Barre, father of Antoine and intendant of Dauphiné.

All of these officials held the title of *conseiller du roi* and had entrance to and a voice in the *conseil d'état*. Unfortunately the title was not restricted to these types of officials, but was widely distributed, if not sold, among many different officeholders including members of Parlement, *secrétaires du roi*, and treasurers. Thus the possession of this title is not a sure indication of the most important administrators who deliberated in the royal council.

45. The seven *secrétaires du roi* were: Nicolas Bigot, father of Jacques and former controller general of the *gabelles;* Jean de Choisy, father of Jean and re-

number employed in the sovereign courts increased only slightly to twenty-nine.[46] Thirteen held fiscal positions as treasurers and tax collectors.[47] Seven occupied petty judicial and administrative positions,

ceiver general of finances in the *generalité* of Caen; Jacques de La Fond, father of Claude; Jacques Le Gendre, father of Paul and former receiver general of the *gabelles;* François Le Vayer, father of René and lieutenant general in Le Mans; and Gédéon Tallemant, father of Gédéon, banker at Bordeaux and *trésorier de l'épargne* of Navarre under Henry IV; and Daniel Voisin, father of Daniel.

46. The twenty-nine members of the sovereign courts consisted of four members of the *cour des aides:* Jean Dyel, father of Jacques Dyel de Miromesnil and president of the court in Rouen; François Villemontée, father of François and president of the court in Paris; Michel Le Tellier, father of the secretary of state for war and councilor in the Parisian court; and Claude Le Tonnelier de Breteuil, father of Louis, and grandfather of François, and *procureur général* of the Parisian *cour des aides.* Three members of the *grand conseil:* Jacques Camus, father of the Camus brothers, was an *avocat* in this court, while two former masters of requests, Louis Machaut, father of Louis, and François Pomereu, father of Auguste-Robert, both became presidents of the *grand conseil.* Five members of the *chambre des comptes* in Paris: Jean Boucherat, father of Louis; Jean Sublet, father of François Sublet de Noyers; Jacques Gobelin, father of Claude; and Louis Machaut, father of Charles; all served as ordinary councilors, or *maîtres.* Jacques Goury, father of Pierre, served as *auditeur* in the court.

Seventeen fathers were judicial officials in the parlements, including eight presidents: Jacques Amelot, father of Jacques Amelot de Beaulieu, was president in a *chambre des enquêtes* of the Parlement of Paris, as was Jean-Jacques Barillon, father of Paul d'Amoncourt. Jean Bochart de Champigny, father of François de Sarron Champigny, had a varied career as master of requests, intendant, and ambassador before ending as a first president of the Parlement of Paris, while Mathieu Molé spent much of his life as a first president of the Parlement in Paris, before ending his last years as a keeper of the seals. Jean-Antoine de Mesmes, father of Jean-Jacques, was a president *à mortier* of the Parisian court. Three fathers were presidents of parlements other than that of Paris: Raoul Bretel de Grémonville, father of Nicholas, was a president *à mortier* of the Parlement of Rouen; Louis Cazet de Vauxtorte, father of François, was a president in the Parlement in Brittany; and Elie Laisné, father of Louis, sieur de la Marguerie, was a former master of requests who became the first president of the Parlement of Aix.

Nine other fathers served in parlements: Claude de Heere, father of Denis; François de Lauzon, father of Jean; and Jérôme Le Maistre, father of Louis, sieur de Bellejamme, served as ordinary *conseillers* in the Parlement of Paris: Antoine Arnauld, father of Robert d'Andilly and grandfather to Simon de Pomponne; Gilles Baussan, father of Philibert; Michel Brachet, father of Jacques and Charles, brother army intendants; and Jean Joüenne, father of René d'Esgrigny, functioned as *avocats* in the Parlement of Paris. Antoine Servien, father of Ennemond, was a councilor in the Parlement of Dijon, while Etienne Pontac, father of Pierre, was *greffier en chef* of the Parlement of Bordeaux.

47. The thirteen treasurers and fiscal officials are: Claude d'Aubray, father of Dreux; Claude Bazin de Bezons, father of Claude; Louis Le Peletier, father of Michel de Souzy; and Antoine Ribeyre, father of Jean, all treasurers of France in the bureaus of finance in Soissons, Châlons, Grenoble, and Riom, respectively. Nicolas Robert, Louis's father, was supposedly a treasurer of France at Riom, according to the Roberts' verification of nobility in 1667, B.N. *nouveau d'Hozier* 286, family "Robert," fols. 29–35, but other genealogical records in B.N. *dossiers bleus* 569, family "Robert (de la Fortelle)," fol. 2, describe Nicolas as only a *commis de l'épargne, secrétaire de la chambre.* Charmeil, *Trésoriers de France,*

Table 3
Occupations of Intendants' Fathers

---

  5 lived nobly at court
  7 *secrétaires du roi*
15 high government officials (councilors of state, masters of requests, intendants)
29 members of sovereign courts, including:
     3 *grand conseil*
     4 *cour des aides*
     5 *chambre des comptes*
   17 parlementarians
13 treasurers and important fiscal officers
  7 lesser judicial and administrative positions
  2 bourgeoisie

---

78 Total

---

mostly in the provinces.[48] Only two are clearly identified as bourgeois.[49]

One noticeable change occurred between the two generations. Although the magistracy of the sovereign courts held its own as a choice of professions (twenty-nine fathers as opposed to twenty-five grandfathers), it had decreased in relation to the total number (twenty-nine out of seventy-eight versus twenty-five out of fifty-eight). But the

pp. 473–74, does not mention Robert in his listing of the treasurers of France at Riom.

Claude Le Roy de la Potherie, Charles's father, was a treasurer of the extraordinary funds for war, while Guillaume de Mouceau, father of another Charles, held the same position but in the province of Béarn. Charles Brèthe, father of Charles Brèthe de Clermont, was a controller general of the same treasury of the extraordinary funds for war.

Louis I Chauvelin's father, Louis, was a receiver of the domains and forests of the generality of Paris and the Ile-de-France, while Nicolas Colbert, parent of Charles Colbert de Croissy, was a receiver of *aides* and paymaster for the *rentes* of the Hôtel de Ville. François Du Gué, Gaspard's father, was an official of the *grenier à sel* at Moulins. Jacques Paget, father of Jacques, was a receiver of *tailles* in Picardy. Finally, Guillaume de Sève, Alexandre's father, was a receiver general and a paymaster for the *rentes* of the clergy.

48. The seven fathers employed in lesser judicial and administrative positions are: Claude Balthazar, Jean's father, who was a councilor in the *présidial* of Sens; Denis Gombault, Louis's father, who held the same position in Troyes; François Lasnier, François's father, who was a lieutenant general and president of the *présidial* of Angers, as well as mayor of that city. Louis Carlier, father of Etienne, was a *procureur* in the *siège royal* of Clermont. Jacques Charuel's father, Jacques, was a clerk in the *bailliage* and *prévôté* of Epernay. Philippe Talon, father of the Talon intendants, was an *avocat* to the *bailli* of Châlons. One other, Jean Heisse, Thomas's father, was a resident of the elector of the Palatinate at the French court and French agent in Germany.

49. Jean Bazin, father of François, was a wealthy merchant draper of Paris, while Denis Marin's father was a humble *savatier et passeur sur la Saône*, according to Charmeil, *Tresoriers de France*, p. 258. A *savatier* is one who repairs old shoes and a *passeur* is a ferryman, perhaps indicating that Marin's father was not very successful in his enterprises.

number of fathers holding the highest office in the sovereign courts, that of president, increased from two to eight. At the same time, there was a significant increase in the number of ancestors who held high government positions (from three to fifteen). A gradual transformation upward had begun, from lower to higher positions in the sovereign courts and from judicial ranks to the upper administrative ranks, anticipating the large influx of men into the higher echelons of government during the lives of the third generation. These men would serve as army intendants as well as in many other positions in the upward course of their careers.

A brief survey of intendants' occupations before their army intendancies indicates that twenty-four out of a total of ninety-three began their careers by service in the Parlement of Paris.[50] Another eight intendants were officials in provincial parlements, although one of them, Louis Le Tonnelier de Breteuil, later shifted to the Parlement of Paris.[51] In all, a total of thirty-two had judicial experience as parlementarians. Another twenty-three had served in the thirty-odd sovereign courts in France, such as the *cours des aides, chambres des comptes,* and the *grand conseil* of Paris.[52] A grand total of fifty-five out of ninety-three

50. Members of the Parlement of Paris included: Jean Balthazar, Paul Barillon d'Amoncourt, Guillaume Bautru de Serran, Antoine Bordeaux, Louis II Chauvelin, Jean de Choisy, Dreux-Louis Du Gué de Bagnols, Claude Gobelin, Denis de Heere, Jean Lauzon de Liré, Antoine Le Fèvre de la Barre, Louis-François Le Fèvre de Caumartin, Olivier Le Fèvre d'Ormesson, Louis Le Maistre de Belle-jamme, Michel Le Peletier de Souzy, Charles Le Roy de La Potherie, François Le Tonnelier de Breteuil, Geoffrey Luillier d'Orgeval, Jean-Jacques de Mesmes, Jean-Edouard Molé de Champlastreux, Gédéon Tallemant, François Villemontée, Daniel Voisin, and René de Voyer d'Argenson.

The number ninety-three refers to the individuals listed in the appendix; it does not include the Talon family, because its members are impossible to distinguish.

51. Jean-Etienne Bouchu, Nicolas Fouquet, and Charles Colbert de Croissy were councilors in the Parlement of Metz; Paul Le Gendre was a *procureur général* in that same parlement. Paul Hay du Châtelet and Louis Le Tonnelier de Breteuil were members of the Parlement of Rennes in Brittany. Honoré Courtin and Jacqus Dyel de Miromesnil were councilors in the Parlement of Rouen.

52. Pierre de Pontac was first president in the *cour des aides* of Bordeaux. Charles Brèthe de Clermont, Pierre Goury, and Jean Ribeyre were *maîtres* in the *chambre des comptes* in Paris, while Ennemond Servien was president in that of Grenoble. Many more intendants began their careers in the *grand conseil:* Sieur Ollier and Jacques Amelot de Beaulieu served as presidents of the *grand conseil,* while Claude Bazin de Bezons and François Cazet de Vauxtorte served as *avocats généraux* to this court. Additionally, Charles Camus du Clos and Isaac Laffemas began their careers as ordinary *avocats* in the court. Finally, Dreux d'Aubray, François Bazin, François Bochart de Sarron Champigny, Nicolas Bretel de Grémonville, Louis I Chauvelin, Louis Laisné de la Marguerie, François Lasnier, Michel Le Tellier, Charles Machaut, Louis Machaut, Auguste-Robert de Pomereu, and Alexandre de Sève served as *conseillers* in the *grand conseil.*

The *grand conseil* was the second most important (after the parlements) train-

army intendants had acquired judicial experience in the highest courts of the land.

There came a critical juncture in every judicial career: a man could either remain a legal officer or make the transition from the courts to the royal council; that is, leave the judicial ranks for a more lucrative position in the royal administration and a chance to become one of the greatest officials in the land. Although the social prestige of the law-yer's long robe was high, even greater awards could be obtained from the crown. But this usually meant beginning as a master of requests, the lowest rung of the administrative hierarchy. The rise to minister fol-lowed this classical pattern: after obtaining a basic education in a *col-lège* and spending some time at a law school, a man acquired a position in the sovereign courts, hopefully in the Parlement of Paris if he had enough money and connections. After a few years of judicial service, he would obtain the position of master of requests and spend the rest of his career in government administration before rising to a secretary-ship of state.[53]

The careers of four ministers illustrate this pattern: Louis Boucherat, Nicolas Fouquet, Charles Colbert de Croissy, and Michel Le Tellier. Louis Boucherat began his climb to the position of chancellor of France in 1637 when he obtained the office of *correcteur* in the Parisian *chambre des comptes*, undoubtedly through the assistance of his father, who also served in that court. Boucherat quickly bettered his situation by becoming a councilor in the Parlement of Paris, and then in 1643 a master of requests. In the royal administration, he served in a variety of roles: as provincial intendant in Guyenne and Langue-doc, as army intendant in the army of Haute Guyenne during the Fronde, as a special commissioner to prepare an inventory of Nicolas Fouquet's papers for his trial, and as a ten-time commissioner sent to negotiate with the estates of Brittany. At last, in 1685, he became chancellor of France and added the position of keeper of the seals six years later.

His rival, Nicolas Fouquet, launched his career as councilor of the Parlement of Metz after his education in the *collège* of Clermont. Acquiring the charge of master of requests in 1636, he served as army

ing ground for future intendants and administrative officials in the seventeenth century, because it judged cases in which the government sought a more favor-able decision than might be expected in the regular courts. The court handled affairs that the crown had withdrawn from the parlements and other courts or matters that were sent to it directly from the royal council. Consequently, bright, ambitious young men might attract the government's attention in this position. Gruder, *Provincial Intendants*, pp. 48–50.

53. Gruder, *Provincial Intendants*, pp. 52–56. Mousnier, *Lettres au Séguier*, 1:56–57.

intendant for a brief period in 1642 for the defense of the northern frontier. In 1643 the government transferred him to Dauphiné as provincial intendant, although he also served as intendant of the troops stationed in that province. By 1647 he was once again on the northern frontier, this time as intendant of the army assembled in Picardy. With the coming of the Fronde, he served as army intendant, this time in Normandy and the region around Paris. In 1650 he obtained the post of *procureur général* of the Parlement of Paris. Three years later he emerged as surintendant of finances and one of the key people in Mazarin's government—a position he retained until Louis XIV's ascension to power and his own imprisonment in 1661.

Charles Colbert de Croissy began his career as an assistant to Le Tellier and served in several minor military expeditions in the 1640s as army intendant. Then he obtained his mastership of requests in 1663 and served successively as intendant of the provinces of Alsace, Touraine, Anjou, Maine, Amiens, and Soissons. With the commencement of the War of Devolution in 1667, he served as army intendant with the royal army, and after the war he obtained the intendancy of Paris as a reward for his services. He then entered the diplomatic service and acted as ambassador to England from 1668 to 1674 and as negotiator at the conference of Nimwegen, before obtaining the presidency *à mortier* of the Parlement of Paris in 1679. With Pomponne's disgrace in 1680, Colbert de Croissy replaced him as secretary of state for foreign affairs, where he remained until his death in 1696.

Michel Le Tellier acquired his first position as a judicial official in the *grand conseil* in 1624, before acting as *procureur du roi* at the court of the Châtelet in 1631. By the end of that decade he had obtained the prerequisite mastership of requests and entered the upper reaches of the royal administration. He served as intendant of the army in Italy in 1640 and attracted the attention of Mazarin. When that minister rose to power in 1643, he named him secretary of state for war. Le Tellier held this position until 1677, when he was rewarded for his past service with the position of chancellor and keeper of the seals—a position he exercised until his death in 1685.[54]

The acquisition of a mastership of requests was the process by which a judicial official, a member of the robe, became a king's man. The office drew him away from his former parlementary colleagues

54. For information on the four officials, see Prevost and Roman d'Amat, *Dictionnaire de biographie française*, s.v. "Boucherat (Louis)"; Jules-Auguste Lair, *Nicholas Fouquet procureur général, surintendant des finances, ministre d'état de Louis XIV* (Paris, 1890), 1:63–99 and passim; André, *Michel Le Tellier et l'armée*, pp. 33–43; and the chapter on Colbert de Croissy in Mousnier, *Conseil du roi*, pp. 153–74.

and molded him into a royal official who completely identified himself with the government and its policies. A mastership of requests had two aspects. As part of the *requêtes de l'hôtel du roi,* a master acted as judge in cases involving members of the royal household and certain other privileged officials, as well as in various matters referred by the royal council. As part of the royal council, the masters of requests were reporters to the different state councils: privy, finances, and direction. They prepared the dossiers for the council sessions and conducted preliminary investigation concerning cases. In this manner the masters of requests had an excellent opportunity to learn about the royal bureaucracy, for their job necessarily led them to consult a wide variety of officials. The crown also assigned the masters of requests to a large number of bureaus and extraordinary commissions that assisted the royal councils in drafting and executing resolutions. The result was that a master of requests was usually an experienced official who had an intimate knowledge of the government's operation.[55]

Of the ninety-three intendants examined, forty-eight had served as masters of requests,[56] thirty-seven had not,[57] and four (Michel Aligre

55. Gruder, *Provincial Intendants,* pp. 52–56, and Marion, *Dictionnaire des institutions de la France,* s.v. "Maîtres des requêtes."

56. The forty-eight masters of requests were Jacques Amelot de Beaulieu, Dreux d'Aubray, Jean Balthazar, Paul Barillon d'Amoncourt, François Bazin, François Bochart de Sarron Champigny, Antoine Bordeaux, Louis Boucherat, Jean-Etienne Bouchu, Jacques de Chaulnes, Louis I Chauvelin, Louis II Chauvelin, Jean de Choisy, Honoré Courtin, Dreux-Louis Du Gué de Bagnols, Jacques Dyel de Miromesnil, Nicolas Fouquet, Claude Gobelin, Nicolas Bretel de Grémonville, Paul Hay du Châtelet, Denis de Herre, Jeannin, Isaac Laffemas, Claude de La Fond, Louis Laisné de la Marguerie, François Lasnier, Jean Lauzon de Liré, Antoine Le Fèvre de la Barre, Louis-François Le Fèvre de Caumartin, Olivier Le Fèvre d'Ormesson, Louis Le Maistre de Bellejamme, Michel Le Peletier de Souzy, Charles Le Roy de La Potherie, Michel Le Tellier, François Le Tonnelier de Breteuil, Louis Le Tonnelier de Breteuil, Geoffrey Luiller d'Orgeval, Charles Machaut, Louis Machaut, Jean-Edouard Molé de Champlastreux, Ollier, Jacques Paget, Auguste-Robert de Pomereu, Alexandre de Sève, Gédéon Tallemant, François Villemontée, Daniel Voisin, and René de Voyer d'Argenson.

57. The thirty-seven army intendants who were not masters of requests were Robert Arnauld d'Andilly, Simon Arnauld de Pomponne, Philibert Baussan, Guillaume Bautru de Serran, Claude Bazin de Bezons, Jacques Bigot, Guillaume Bordeaux, Charles Brachet, Louis Brachet, Charles Brèthe de Clermont, Germain-Michel Camus de Beaulieu, Charles Camus du Clos, Etienne Carlier, François Cazet de Vauxtorte, Jacques Charuel, Gilbert Colbert de Saint Pouenges, Jean Colbert du Terron, the sieur de la Croix, Gaspard Du Gué, Pierre Gargan, Louis Gombault, Pierre Goury, Thomas Heisse, Pierre Imbert, René Joüenne d'Esgrigny, Jacques-Louis Le Marié, René Le Vayer, Michel-Louis Malézieu, Denis Marin, Charles Mouceau de Nollant, Julien Pietre, Pierre de Pontac, Jean Ribeyre, Louis Robert, Ennemond Servien, François Sublet de Noyers, and René de Voyer de Paulmy (Dorée). Information was unavailable on these army intendants: D'Estampes de Valençay, the sieur de la Court, Le Goux de la Berchère, the Talons, and Raymond Trobat.

Cf. Mousnier's comments, *Lettres au Séguier,* 1:57, that "une infime minorité"

de Saint-Lié, Charles Colbert de Croissy, Paul Le Gendre, and Jean-Jacques de Mesmes) became masters of requests after serving as army intendants. Information on the rest was unavailable.

There are various explanations for the number of non–masters of requests who served as army intendants. Some, such as René de Voyer de Paulmy, known as Sieur Dorée and a cousin of Argenson, was a logical substitute intendant in the absence of his relative. Other local officials, such as René Le Vayer, president of the sovereign council of Arras, and Julien Pietre, treasurer of France at Amiens, were obvious choices for temporary army intendancies in their regions during the incapacity of the regular intendants. Men living near frontier regions and well acquainted with local conditions also made good selections for a military intendancy. For example, Ennemond Servien, native of Dauphiné, president of the sovereign council of Pignerol, and ambassador to Savoy, was obviously fitted for an army intendancy in Italy. The fact that his brother, Abel Servien, was secretary of state for war from 1630 to 1636 and then surintendant of finances did not hurt his chances either. Other men were army intendants in time of emergency, particularly during the Fronde when loyalty counted more than experience as a master of requests. During this period, regular Parisian intendants of finance such as Pierre Gargan and Denis Marin, and members of the different sovereign courts such as Pierre de Pontac, Jean Ribeyre, and Jean-Jacques de Mesmes, faithfully performed the functions of army intendant. One army intendant, Pierre Imbert, had past experience provisioning armies as a *commissaire général des guerres*—an office that once was a rival of that of army intendant. Other men, such as Jacques Bigot, controller general of the infantry and of the extraordinary funds for war, had expertise in military finances.

The explanation for most of the non–masters of requests is that they were part of a clientage system of friends and relatives of some powerful patron. Many of the army intendants—and even provincial intendants—fit this pattern during the early years of the Le Tellier administration. Philibert Baussan's early experience as a legal official in the Châtelet did not qualify him as provincial intendant of Alsace or of the armies in this region as much as did the fact that he was Le Tellier's relative. The same is true for another of Le Tellier's relatives, Charles Brèthe de Clermont. One army intendant, a Talon, owed his position

of non–masters of requests became provincial intendants during Séguier's chancellorship. Mousnier is in error in claiming that Isaac Laffemas was not a master of requests; he was, being received in his office in 1627. See Georges Mongrédien, *Le Bourreau du cardinal de Richelieu, Isaac de Laffemas* (Paris, 1929), p. 54, and B.N. *dossiers bleus* 375, family "Laffemas."

as army intendant in Catalonia and Champagne to his collaboration with Le Tellier during the latter's own intendancy in Italy.[58]

Le Tellier quickly recognized, however, that good administrators had to have expertise as well as devotion. By the 1660s he had perfected an apprenticeship system for the selection and training of army intendants. Relatives, such as Louis Robert and the sieur de la Croix, and clients, such as Jacques Charuel, Etienne Carlier, and the Camus brothers, began as *commis*, clerks, in the war department or as *commissaires des guerres*. They were groomed for important military intendancies through a careful process of selection and experience, progressing from missions with minor expeditions to service in large armies.

Louvois took over his father's system and increasingly used his *commissaires des guerres* as army intendants. Men such as Thomas Heisse, Rene Joüenne d'Esgrigny, Michel-Louis Malézieu, Jacques-Louis Le Marié, and Charles Mouceau de Nollant were drawn from these ranks and spent their lives in military service. One might add the names of other Louvois clients such as Louis d'Amorezan and Jacques de La Grange, who did not serve as army intendants but were *commissaires des guerres* before becoming provincial intendants of frontier regions like Hainaut and Alsace. It is clear that by the latter half of the seventeenth century the office of *commissaire des guerres* might substitute for a mastership of requests in military administration and be a stepping-stone to higher offices in the civilian government.

This conclusion is obvious if one examines the twenty-seven army intendants who served after 1660. Thirteen were drawn from the ranks of the provincial intendants, men who had entered the royal service as masters of requests.[59] The fourteen other men had served as *commissaires des guerres* or *commis* in the war department under Le Tellier and Louvois before their appointment as army intendants.[60] These fig-

58. André, *Michel Le Tellier et l'armée*, pp. 635–36.

59. Paul Barillon d'Amoncourt, François Bazin, Jean-Etienne Bouchu, Louis II Chauvelin, Charles Colbert de Croissy, Honoré Courtin, Dreux-Louis Du Gué de Bagnols, Claude de La Fond, Louis-François Le Fèvre de Caumartin, Michel Le Peletier de Souzy, Louis Machaut, and Auguste-Robert de Pomereu. Raymond Trobat, provincial intendant of Roussillon, has been added to this list, but I could not find evidence that he was ever a master of requests; he seems to have been a local official in Roussillon who was placed in the intendancy because of his knowledge of the region.

60. Germain-Michel Camus de Beaulieu, Charles Camus du Clos, Etienne Carlier, Jacques Charuel, Gilbert Colbert de Saint Pouenges, the sieur de la Croix, Thomas Heisse, René Joüenne d'Esgrigny, Jacques-Louis Le Marié, Michel-Louis Malézieu, Charles Mouceau de Nollant, and Louis Robert. Jean Colbert du Terron has been included in this group because he served as army intendant for the troops sent to Messina in Italy in 1675. He began his career as a *commis* in the war department under Le Tellier and served in several army intendancies in the 1650s before turning to naval administration and acting as head of the French marine until 1673. His intendancy with the expedition to Messina was due to special circumstances,

ures are in agreement with what is known about eighteenth-century military administration, namely that army intendants were regularly drawn from two groups: provincial intendants on the frontier who were closest to the army, and the *commissaires des guerres*.[61]

The major difference between the two types of administrators lay in their education. Officials who went the member of parlement–master of requests route had to have a legal education, while those who rose through the ranks of military administration normally did not have this background. Entrance into the sovereign courts and a mastership of requests required a *licencié ès lois*, certification normally gained after three years in a law school. Of course loopholes always existed in the ancien régime for those who had the wealth and social position to take advantage of them. Thus legal studies were often rapid and superficial; students could hire others to attend lectures and take exams for them. The quality of instruction, particularly in the provincial law schools, was remarkably low. Nevertheless, the intendants belonged to the educated elite of the society.[62]

A further question remains about the heights to which the ex–army intendants rose in the king's service. Only a few, such as Simon Arnauld de Pomponne, Louis Boucherat, Charles Colbert de Croissy, Nicolas Fouquet, Michel Le Tellier, and François Sublet de Noyers, ascended to the very top as secretaries of state, surintendants of finance, or chancellors. Five other men obtained positions as presidents of the sovereign courts, but most former intendants ended their careers in the council of state.[63] Indeed, the normal career for a master of requests was service in that position for six to ten years, resignation, and then a move up as

---

and he cannot be considered a normal army intendant during the 1660s. He was not a master of requests, but he owed his position first to the patronage of Le Tellier and then to his relative Jean-Baptiste Colbert. Similarly, the Talon who served as army intendant in Flanders in 1660 should probably be included among the non–masters of requests. Although I could not precisely identify which member of the Talon family this was, not one of the family was a master of requests though several were *commissaires des guerres*. B.N. *cabinet d'Hozier* 316, family "Talon," and B.N. *collection de Chérin* 192, family "Talon." This family was distantly related to the famous Omer Talon.

61. See the comments of Antoine de Pas, marquis de Feuquières, *Mémoires de M. le marquis de Feuquiere, lieutenant général des armées du roi; contenans ses maximes sur la guerre, et l'application des exemples aux maximes*, new ed., rev. and cor. (London, 1736), p. 58; and François Sicard, *Histoire des institutions militaires des Français* (Paris, 1834), 1:416.

62. Bluche, *Magistrats du parlement*, pp. 59–62. Cf. the remarks of Gruder, *Provincial Intendants*, pp. 23–34, which are much more optimistic about the legal education obtained in the eighteenth century.

63. Jacques Amelot de Beaulieu as first president of the Parisian *cour des aides*, Pierre de Pontac as first president of the *cour des aides* in Guyenne, Jean-Jacques de Mesmes and Jean-Edouard Molé de Champlastreux as presidents *à mortier* in the Parlement of Paris, and Louis Robert as president of the Parisian *chambre des comptes*.

conseiller d'état. Many other officials in the ancien régime also obtained the title, but only a select few actually received a brevet to enter the council with a deliberative voice, first on a semester, or six-month basis, then as regular councilors, *conseillers d'état ordinaires*.[64] Although evidence is unavailable for all the masters of requests who served sometime in their career as army intendants, at least eighteen, and probably many more, obtained this position as ordinary councilor.[65] Another seven former army intendants, who do not appear to have been masters of requests, obtained this office too.[66] Two additional masters of requests finished their service in a special division of the royal council known as the *conseil royal des finances*, Michel Aligre de Saint-Lié and Auguste-Robert de Pomereu.

One master of requests, Jacques Paget, retained his office for the rest of his life, rising through seniority to become *doyen*, while another, Denis de Heere, resigned his mastership only to die the same year. Another ten army intendants, mostly masters of requests, saw diplomatic service as ambassadors.[67] One intendant, Jean Lauzon de Liré, served as governor of Canada; two others ended up in the military sphere: Antoine Le Fèvre de la Barre as a lieutenant general in the royal armies, and Michel Le Peletier de Souzy as director general of fortifications— an office separated from the war department at Louvois's death in 1691. Another three held important positions at court: Guillaume Bautru de Serran as chancellor to Louis XIV's brother, Philippe; Jean de Choisy as chancellor to Gaston, brother of Louis XIII; and Gilbert Colbert de Saint Pouenges as grand treasurer of the king's orders. Two ex-intendants, Dreux d'Aubray and Isaac Laffemas, obtained the important Parisian judicial office of lieutenant civil of the Châtelet, although Laffemas was disgraced after the death of his patron Richelieu. One army intendant, Robert Arnauld d'Andilly, retired to solitude at Port Royal,

64. Mousnier, *Lettres au Séguier*, 1:58–59.

65. Paul Barillon d'Amoncourt, Guillaume Bautru de Serran, François Bochart de Sarron Champigny, Jean-Etienne Bouchu, Louis II Chauvelin, Honoré Courtin, Jacques Dyel de Miromesnil, Dreux-Louis Du Gué de Bagnols, Claude Gobelin, Louis Laisné de la Marguerie, Louis-François Le Fèvre de Caumartin, Olivier Le Fèvre d'Ormesson, Louis Le Maistre de Bellejamme, Charles Le Roy de La Potherie, Louis Le Tonnelier de Breteuil, Charles Machaut, Alexandre de Sève, and René de Voyer d'Argenson.

66. Philibert Baussan, Claude Bazin de Bezons, Charles Brèthe de Clermont, Guillaume Bordeaux, Jean Colbert du Terron, René Le Vayer, and Denis Marin.

67. Paul Barillon d'Amoncourt, François Bazin, Antoine Bordeaux, Nicolas Bretel de Grémonville, François Cazet de Vauxtorte, Honoré Courtin, François Lasnier, Paul Le Gendre, Ennemond Servien, and René de Voyer d'Argenson. One should not forget that at least three army intendants died in service: Louis I Chauvelin in the army of Piedmont in 1645, Jacques Brachet in the army of Italy in 1659, and the sieur de la Croix on the expedition to Crete in 1669.

but he was at the end of his career anyway. Perhaps the most startling metamorphosis was that of François Villemontée, who felt a call to the religious life, obtained a separation from his wife, and became a priest and later bishop of Saint Malô.

The fate of various clients of the Le Tellier family who served as assistants or *commissaires des guerres* in the war department is interesting. These "new men" did not spend the rest of their lives in military service, but transferred to civilian administration. Instead of a mastership of requests, they used their relationship with the Le Tellier family to rise above their army intendancies. After the Dutch War, Louis Robert purchased the office of president of one of the sovereign courts in Paris, the *chambre des comptes*. Etienne Carlier became a provincial intendant in Roussillon, while Jacques Charuel spent the rest of his career, after his last army intendancy in 1673, as provincial intendant of Lorraine and in the three bishoprics of Metz, Toul, and Verdun. René Joüenne d'Esgrigny served as a provincial intendant in Catalonia in 1695, while Michel-Louis Malézieu served as intendant along the Champagne frontier. The Camus family passed the office of controller general of the artillery among its members as a family possession, although one of them (Germain-Michel Camus de Beaulieu) also served as provincial intendant in Roussillon.

This analysis of social mobility can be concluded by examining the sons of the army intendants. As many sons as fathers rose to the highest position in the state, in part because several fathers obtained the *survivance* or right of succession for their offspring. Thus Louvois succeeded his father as secretary of state for war, as did Jean-Baptiste Colbert, marquis de Torcy (Colbert de Croissy's son), the secretary of state for foreign affairs. Michel Le Peletier de Souzy's son, Michel, served as controller general of finances from 1726 to 1730 with the rank of minister of state. Robert Arnauld d'Andilly's son, Simon Arnauld de Pomponne, became a secretary of state for foreign affairs, but he had already been an army intendant in his own career and probably should be counted among the intendants rather than their sons. Louis II Chauvelin's son, Germain-Louis, was keeper of the seals, minister, and secretary of state for foreign affairs in 1727; François Le Tonnelier de Breteuil's son, François-Victor, served two terms as secretary of state for war, once from 1723 to 1726 and again from 1740 to 1743. Although René de Voyer d'Argenson did not have a son in the ministerial ranks, he did have two grandsons: Marc-Pierre, count d'Argenson, the secretary of state for war from 1743 to 1757, and his brother, René-Louis, marquis d'Argenson, the secretary of state for foreign affairs from 1744 to 1747.

Table 4
Occupations of Intendants' Sons

12	lived nobly, including:
	6 court positions
	6 knights of Malta
6	ministers and secretaries of state
23	high government officials (councilors of state, masters of requests, intendants)
21	members of sovereign courts
8	lesser fiscal, judicial, and administrative positions
26	army and navy officers
25	ecclesiastical careers, including:
	2 archbishops
	4 bishops

121 Total

Very few families rose to this exalted height. Most settled for establishing their sons in the robe as officeholders in the sovereign courts, in the royal administration as masters of requests and intendants, or in the nobility of the sword as army officers or members at court. Information is available on one hundred fifteen other descendants of army intendants who held positions below that of secretary of state. Twenty-one held positions in the sovereign courts, including seven presidencies: *à mortier* in the Parlement of Paris (Mesmes and Molé), of the *chambres des enquêtes* in the Parlement of Paris (Le Maistre and Amelot), the first president of the Parlement of Aix (Marin), a president of the Parisian *cour des aides* (Amelot), and one of the Parisian *chambre des comptes* (Du Gué de Bagnols). Twenty-three served as masters of requests, intendants, or ambassadors, while eight occupied lesser fiscal, judicial, and administrative positions in the royal administration. Obviously the intendants' offspring sought the most important judicial and administrative positions in the state—a quest in which they were largely successful.

Much more interesting are the sons who attempted to pass into the nobility of the sword. At least three held offices at court, as gentlemen serving the king (the two Gombault sons) and treasurer of the king's orders (Louis II Chauvelin's son), while another three lived nobly: Nicolas-Simon Arnauld as marquis de Pomponne, Louis-Nicolas Fouquet as count of Vaux and viscount of Melun, and Pierre de Voyer as viscount d'Argenson and *grand bailli* of Touraine. Six, mostly younger sons, were enrolled as knights of Malta, which, in principle, required proof of sixteen quarters of nobility for admission, although in practice it was usually sufficient to be from a family that had been noble for a hundred years preceding admission.[68] Twenty-three others served as army officers, and three were captains of vessels in the navy. Of the twenty-three army officers, only five held the highest rank in the army: Jacques, son of Claude Bazin de Bezons, as marshal of France; Louis-François, son of Charles Colbert de Croissy, and Louis Fouquet, marquis de Belle-Isle, as lieutenant generals; and François-Gilbert Colbert, marquis de Saint Pouenges, and Louis Le Tonnelier de Breteuil as *maréchals de camp*. The rest occupied lower rank, mostly as captains of regiments. Nevertheless, one might conclude that officership in the army was both a means to and a witness of social mobility; it probably contributed as much as family alliances to joining the older nobility.

Finally, a number of sons chose ecclesiastical careers. Two rose to be archbishops: Armand, son of Claude Bazin de Bezons, was archbishop of Bordeaux, then Rouen, while Louvois's brother was archbishop of Rheims. Four others were bishops: a Bochart de Champigny of Clermont, a Colbert de Croissy of Montpellier, a Le Tonnelier de Breteuil of Boulogne, and a Sève of Arras. Another nineteen occupied lesser benefices, including three deans, two canons, and the remainder as priors, abbés, and priests.

After examining the wealth of the army intendants, their families, their careers, and their social mobility, one further aspect needs to be examined: their role in seventeenth-century clientage systems. The role of patronage in French politics of the ancien régime has been one of the fascinating problems in recent historical investigation. Although few, if any, of the army intendants applied the term "creature" to themselves, the term is appropriate, for they were aware of the importance of having powerful patrons. Unfortunately, only detailed study can reveal the outlines of a clientage system, for its links were personal ties, quickly forgotten by succeeding generations. It is known that Isaac Laffemas and Charles Machaut—nicknamed *coupeteste* by his contemporaries—were Richelieu's agents. Mazarin had his own clients, including Colbert, Le Tellier, and Fouquet—all of whom rose to positions of importance—but he also had lesser men such as Jean Balthazar. Although Balthazar served incompetently as intendant of

---

68. For entrance into the knights of Malta, see Bluche, *Magistrats du parlement*, p. 322n110. André Corvisier, "Les Généraux de Louis XIV et leur origine sociale," *XVIIᵉ siècle*, nos. 42–43 (1959), p. 49. It should be noted that Roland Mousnier disagrees that there was much fusion between the older nobility and the nobility of the robe. To him, the fact that the children of the administrators sought careers in the sword and the church is explained by the limited number of offices in government and the necessary choice of careers elsewhere. Mousnier also remarks that even when the sons entered the military world, they did not rise to the upper ranks, but remained in the lower grades. Mousnier, *État et société en France aux XVIIᵉ et XVIIIᵉ siècles*, p. 26.

Languedoc until 1647 and after that date as intendant of the army of Italy, he was not disgraced, despite Colbert's accusation of *friponneries*, because of his closeness to the cardinal. The cardinal's affection for his creatures is also illustrated by an incident that occurred when Jacques Brachet died performing his duties as intendant of the army of Italy in 1659. In a letter, Mazarin mourned that "the news of poor Brachet's death touched me deeply, because he was capable, faithful, and devoted." [69]

Much more information is available on the Le Tellier family's clientage system. It included such men as Charles Brèthe de Clermont, Pierre Goury, the Chauvelin family, Claude Bazin de Bezons, the sieur de la Croix, Louis Robert, Etienne Carlier, Jacques Charuel, Gilbert Colbert de Saint Pouenges, Michel Le Peletier de Souzy, and the Camus brothers. Like every patronage system, it was based upon the premise that the clients gave their loyalty and talent to the Le Tellier family in return for advancement and a wide variety of pensions, gifts, lands, and favors. But self-interest was not the only bond that kept the arrangement working—ties of blood and marriage cemented the system. Pierre Goubert's description of French government in the seventeenth century as an "immense ministerial family . . . who placed its nephews in the intendancies, prelacies, and abbeys" is apt. [70]

For example, Pierre Goury, intendant of the army of Catalonia, was a relative of Le Tellier's wife, while Charles Brèthe de Clermont, sometime intendant of Picardy and Artois as well as the army in the 1640s, was Le Tellier's cousin. Louis I Chauvelin, who died while serving as army intendant in Italy in 1645, was another cousin, his father being a brother of Le Tellier's mother, Claude. And the Le Tellier family also protected Louis Chauvelin's son, Louis II, who served as a provincial intendant of the Franche Comté and army intendant in the 1680s. The father of intendant Gilbert Colbert de Saint Pouenges married Le Tellier's sister Claude, while Michel Le Peletier de Souzy's grandfather Jean had married Madeleine Chauvelin, a sister of Le Tellier's maternal

69. On the role of clientage in French government, see Ranum, *Richelieu and the Councillors of Louis XIII*, chaps. 1–2, as well as the general remarks of Mousnier, *Etat et société en France aux XVIIᵉ et XVIIIᵉ siècles*, pp. 16–18. On the reputation of Laffemas and Machaut, see Mousnier, *Lettres au Séguier*, 2:1203–4, 1216–17. Balthazar's difficulties in Languedoc are recorded in Abbé Henry, *François Bosquet, intendant de Guyenne et de Languedoc, évêque de Lodève et de Montpellier: Etude sur une administration civile et ecclésiastique au XVIIᵉ siècle* (Paris, 1889). For Colbert's denunciation of Balthazar, see Jean-Baptiste Colbert, *Lettres, instructions et mémoires de Colbert*, ed. Pierre Clément (Paris, 1861–82), 1:66, 71, 78. Mazarin's letter lamenting Brachet's death is printed in Chéruel, *Lettres du Cardinal Mazarin*, 9:308–10.

70. Pierre Goubert, *Louis XIV et vingt millions de Français* (Paris, 1966), p. 214.

grandfather, Alexandre Chauvelin, making Michel a cousin twice removed of both the Chauvelin and Le Tellier families. Louis Robert's grandmother Marie Chauvelin, a sister of Le Tellier's mother, had married Philibert Choart by her first marriage and gave birth to Robert's mother, thus linking the two families. Philibert Baussan, intendant of Alsace, was a son of Gilles Baussan, who had married the same Marie Chauvelin, Philibert Choart's widow, allying himself with the Robert, Le Tellier, and Chauvelin families. François Du Gué, provincial intendant in Normandy, Lyon, and Dauphiné, married Marie-Angelique Turpin, a sister of Le Tellier's wife. Although François did not serve as army intendant, his uncle Gaspard Du Gué, sieur de Bagnols, did.

Compounding these marriage ties, Gaspard's grandson, Dreux-Louis, provincial and army intendant in Flanders during Louvois's administration, married the daughter of the above François Du Gué and Marie-Angelique Turpin, his own cousin twice removed, thus making himself a distant relative of the Le Tellier family through marriage. The sieur de la Croix, intendant on the ill-fated expedition to Crete in 1669, was another of Le Tellier's kinsmen. Claude Bazin de Bezons, who served as provincial intendant in Languedoc and army intendant in Catalonia, was grandson of Suzanne Choart de Buzenval, into which family Marie Chauvelin, sister of Le Tellier's mother, had married, thus making Claude a distant relative of both the Le Tellier and Robert families.[71]

Sometimes the relationship with the Le Tellier family was so minor that it is difficult for the historian to imagine a claim of parentage. Charles Mouceau de Nollant, army intendant under Louvois, had a great-grandfather Christophe Chauvelin, a brother of Le Tellier's mother's grandfather. It is unknown whether he attracted Louvois's attention by claiming to be a distant relative, but in a large extended family situation anything was possible.[72]

Of course there were other army intendants who were not bound by ties of blood to the Le Tellier family, such as Charuel, Carlier, and the Camus brothers, but they had other bonds. Jacques Charuel's father was a receiver of the lands and seigniory of Louvois in Epernay, and according to the genealogical records in the Bibliothèque Na-

71. André, *Michel Le Tellier et l'armée*, pp. 636–37, and *Le Tellier et Louvois*, pp. 416–18; B.N. *carrés d'Hozier* 180, family "Chauvelin," fol. 230; B.N. *dossiers bleus* 66, family "Baussan," fols. 7, 12–13; 69, family "Bazin," fols. 1, 22, 24; 336, family "Gué, Du," fols. 10–11; 569, family "Robert," fols. 2–3; 627, family "Le Tellier," fols. 9–25, 46–53. Louis André also included genealogical tables of the Chauvelin, Le Tellier, and Le Peletier families in his edition of *Deux Mémoires historiques de Claude Le Peletier* (Paris, 1906), inserted between pp. 46 and 47.

72. B.N. *dossiers bleus* 476, family "Mouceau (Du)," fol. 10.

tionale, his son served as Louvois's secretary before becoming an army intendant. Etienne Carlier had a brother, Pierre, who was a *commis* in the war department and might have attracted Le Tellier's attention through this connection, although it is impossible to prove. The connection of the Camus family with Le Tellier is unknown, but the brothers Germain-Michel Camus de Bealieu, Charles Camus du Clos, and Jacques Camus des Touches worked as *commissaires* in the war department. The Camus family, in turn, cemented its relationship with another client family, that of Mouceau. The Camus brothers' mother was Marie Mouceau, sister of Guillaume, father of army intendant Charles Mouceau de Nollant. This Charles reinforced the alliance by marrying Marie-Charlotte Camus, daughter of Jacques Camus des Touches.[73]

These contorted genealogies prove that the bulk of the Le Tellier clients came from the extended family or families connected to the Le Telliers by service. From this mass of manpower the secretary of war took the men he needed, groomed them as intendants, and bound them to the Le Tellier family by a sense of loyalty. These intendants, in turn, fostered this sense of clannishness by intermarriage among themselves. The result was a unique product, a man bound by ties of blood and devotion to the Le Tellier clan with a strong sense of esprit de corps.

These clients were not reticent about demanding favors in return for their service, and the secretary of war was continually bombarded by requests for favors. Selected correspondence from 1672 and 1673 preserved in the war archives illustrates this boldness: Charuel thanked Louvois for the appointment of his nephew as a lieutenant in the regiment of Picardy, Robert thanked Louvois for giving the abbey of Daimpon to his brother, and Michel Le Peletier asked Le Tellier's protection for the sieur de Corcelles—"my relative," a new *commissaire des guerres* who had just reported his first review of troops. Etienne Carlier asked that a Benedictine abbey in Roussillon be given to his son, Dom Georges, and he did not hesitate to mention possible vacancies to select from. Finally, Charuel requested the position of *aide-major* in the city of Ath for a relative. In this manner the secretary bound the devotion of an entire family to his person, and created an obligation to return the favor by personal services.[74]

Although such gifts officially came from the king, they resulted

---

73. B.N. *dossiers bleus* 150, family "Camus," fols. 2, 11; 171, "Charuel," fols. 2–13; 154, family "Carlier," fols. 4–5.

74. Charuel to Louvois, 17 April 1672, A.G. A[1] 292, fol. 421; Robert to Louvois, 23 August 1672, A.G. A[1] 294, fol. 170; Le Peletier to Le Tellier, 3 July 1672, A.G. A[1] 293, fol. 346; Carlier to Louvois, 10 December 1672, A.G. A[1] 296, fol. 216; Charuel to Louvois, 25 June 1673, A.G. A[1] 350, fol. 323.

from the intermediary actions of the Le Tellier family, who approached the crown for the favors. Other patrons also intervened for their clients, and it was important to have a powerful figure at court always ready to press for his agents. Nothing can illustrate this better than an incident in 1661 when François Bochart de Sarron Champigny requested the abbey of Val Chrestien for his son. Le Tellier replied that several other persons desired the same benefice because of its proximity to Paris. In the end, one of the officers of Mazarin's household obtained the abbey, but Le Tellier promised Bochart that "this will not prevent me from employing my little power [*mon petit pouvoir*] to point out your services and that of your son to His Eminence." Another time Louvois assured Honoré Courtin that "I will render all the service that I can for the son of Marsin because of my affection for you [*pour l'amour de vous*]." [75]

There is no better example of the wheeling and dealing in "the corridors of power" than the case of Claude Bazin de Bezons in 1660–61. In September 1660 that intendant of Languedoc and veteran army intendant requested a change in intendancies. Le Tellier approached Cardinal Mazarin in his behalf, but was rejected. The secretary of war recounted to the disappointed Bezons: "His Eminence honored me by telling me two days ago, on the occasion of M. de Villemontée's consecration, that he knew that I had spoken of the intendancy of Soissons for you, but because you are more useful for the king's service in Languedoc, he had decided that he could not produce it for you." [76]

The scene merits the touch of a novelist's imagination, and perhaps I might be excused for yielding to this temptation. François Villemontée had been a loyal intendant for many years, both in the provinces and the army, until he felt a call to the religious life and became a priest. Mazarin granted him the bishopric of Saint Malô as a reward for his past services. On the day of pomp and solemnity surrounding his consecration, symbolizing the reward of a faithful servant, Mazarin whispered an aside to Le Tellier refusing his request for another client, not through lack of merit but for the benefit of the king's service. Here is the epitome of the clientage system: having a friend in high places to push for you, but having a patron realistic enough to assign favors for the benefit of the king's service. [77]

Bezons had similar difficulties obtaining an abbey for his son. In

75. Le Tellier to Sarron Champigny, 5 February 1661, A.G. A¹ 168, fol. 110; Louvois to Courtin, 25 May 1673, A.G. A¹ 338, fol. 179.
76. Le Tellier to Bezons, 24 September 1660, A.G. A¹ 163, fol. 78.
77. Villemontée's story is recounted in Frédéric Saulnier, *Un Prélat au XVIIᵉ siècle: François de Villemontée, évêque de Saint-Malô (1660–1670), sa femme et ses enfants,* Extrait des mémoires de la Société archéologique d'Ille-et-Vilaine (Rennes, 1903).

October 1660 Le Tellier notified the intendant that he had placed his request before Mazarin and that the cardinal had assured the secretary he would attempt to obtain the desired abbey. A month later Le Tellier again wrote to Bezons, mentioning his efforts on his behalf and that the cardinal had replied that he already had ordered the necessary papers to be prepared and the abbey would be given at a suitable time. But the affair dragged on. In December of that year the cardinal's illness prevented him from working, so Le Tellier could not say anything new about the benefice. In February 1661 Le Tellier remarked to the patient intendant that the time was not ripe for obtaining the abbey, and tried to comfort him: "You must continue as you have begun, not doubting that if this benefice escapes you, you will be satisfied at another occasion." The rest of Le Tellier's letter expressed concern over the cardinal's deteriorating health. In early March, Bezons once more importuned the secretary about the long-sought abbey. Again the secretary disappointed him. He had not delivered Bezons's request, because the cardinal's health prevented him from reading it, but Bezons should remain assured that the cardinal esteemed his person and would certainly procure advantages for his family. Unfortunately for Bezons, the cardinal died shortly afterward and here the correspondence ended. Yet incomplete as the documents are, they are important because they illustrate the patronage process in action and the need to present one's petition at every favorable opportunity.[78]

Every intendant's concern was to remain in his patron's favor, for he constantly needed his master's intervention in his behalf. If someone close to the secretary poisoned his mind against the intendant, that intendant's career could be jeopardized by malicious rumors. Three examples suffice to illustrate this problem, although they are from provincial rather than army intendancies. In 1660 Le Tellier wrote to Antoine Le Fèvre de la Barre, intendant at Riom, confirming the intendant's suspicion that his enemies were slandering him in Paris. The bishop of Clermont, Le Tellier wrote, "has omitted nothing since he has arrived here to put evil in the mind of Their Majesties and His Eminence in your regard." Fortunately for the intendant, the crown was convinced of his integrity and that his only crime was zeal in recovering back taxes; but Le Tellier warned him, "as your servant, I am obliged to tell you that it is desirable that you change your manner."[79]

---

78. Le Tellier to Bezons, 15 October 1660, 5 November 1660, 24 December 1660, 5 February 1661, and 4 March 1661, A.G. A¹ 163, fols. 129, 204, 364, and A¹ 168, fols. 113, 153. Cardinal Mazarin died on 9 March 1661.
79. Le Tellier to La Barre, 28 November 1660, A.G. A¹ 163, fol. 271.

In 1670 the intendant of the three bishoprics of Metz, Toul, and Verdun, Jean-Paul de Choisy, was equally concerned about his position, for he believed that army intendant Gilbert Colbert de Saint Pouenges had denounced him for imagined faults in not providing assistance during the Lorraine campaign. In his letter, Choisy stressed that he knew Saint Pouenges was Louvois's relative and was in a position to harm him, but Louvois reassured Choisy that he was mistaken, for Saint Pouenges had not spoken against him; on the contrary, he had praised him. Louvois frankly added that if Choisy had further doubts, he should feel free to voice them.[80]

Finally, in 1672 spiteful tongues accused Michel Le Peletier de Souzy of not paying proper attention to the wounded soldiers at the hospital of Deins. The intendant wrote to Louvois justifying his conduct and remarked: "I desire with all my heart that those who will have the intention of rendering me bad offices near you will always choose their field of battle as poorly as they have done this time." Le Peletier must have satisfied all doubts, for he remained intendant of Flanders until 1683.[81]

These three examples underline the importance of remaining in the good graces of one's patron, although the Le Tellier family did not put too much stock in idle rumors. The essential part of any clientage system remained a fundamental trust between patron and client, and over the years close personal ties developed between the secretary and his creatures, even with such a strong-tempered person as Louvois. A good example of the warmth that developed between a sometimes brutal master and his subordinate is Jacques Charuel. Charuel had never married, and his gratitude to Louvois was so strong that he willed his entire fortune, a sum in excess of 350,000 *livres*, to the secretary, but Louvois died three months before the intendant, forcing Charuel to destroy his will. Incidents such as this are a reminder that the clientage system was not always based upon cold calculation for personal profit, but upon admiration and respect by both parties.[82]

80. Choisy to Louvois, 3 November 1670, A.G. A¹ 250, pp. 205*bis*v–7v; reply of Louvois, 20 November 1670, A.G. A¹ 252, pp. 140r–41r.
81. Le Peletier to Louvois, 11 August 1672, A.G. A¹ 294, fol. 67.
82. B.N. *dossiers bleus* 171, family "Charuel," fols. 2–13.

III

Functions of the Early Army
Intendants in the 1630s

The functions of the early army intendants are difficult to determine, since the only documents remaining are partial drafts of commissions and a few scattered letters buried among various archives. One must not rely too heavily upon the general phrases in these commissions, for they are at best only statements of intent. Institutional history cannot be written merely from the ordinances of the ancien régime; historians ask whether officials really functioned that way. Here the fragments of preserved correspondence are important to give flesh to dry bones. Nevertheless, the commissions of the army intendants are important as statements of government policy. Although there were variations, commissions had certain features in common in the 1630s. This standardization persisted throughout the seventeenth century and is the logical beginning for any study of an army intendant's functions.[1]

The formula was as follows. First, the royal salutation: "Louis, by the grace of God, king of France and Navarre, to our beloved and faithful ———— [the official's last name or title and the offices he possessed], greetings." In the 1630s the wide variety of commissioners had narrowed down to men who were *conseillers d'état* and possessed a mastership of requests—although occasionally other important officials

---

1. Army intendants also received a detailed set of instructions with their commissions. Unfortunately, very few of these remain, and I could not locate any before Le Tellier's administration in the 1640s; these are discussed in the next chapter. On the dangers of writing administrative history from royal edicts, see Orest A. Ranum's cogent remarks, *Richelieu and the Councillors of Louis XIII: A Study of the Secretaries of State and Superintendents of Finance in the Ministry of Richelieu, 1635–1642* (Oxford, 1963), p. 46.

were appointed, such as Gaspard Du Gué, treasurer of France in the bureau of finance at Lyon.[2]

After the salutation came a short preamble or explanation giving the rationale behind the nomination. For example, the commission might read: "Having resolved to assemble in our province of Champagne a powerful army . . . to oppose those of our enemies who wish to threaten our realm," or "having resolved to enter Italy with a powerful army . . . to guarantee our neighbors and allies from invasion . . . and to oppose the designs of those who would threaten our authority," or because of "our desire that all our people live in the repose that is so necessary and to suppress the troubles caused by our disaffected subjects."[3]

Next, the commission explained in general terms the need for an intendant. Sometimes it was to assist a general. The commission of René de Voyer d'Argenson, dated 12 August 1632, declared: "Judging how important it is for his [Condé's] assistance to place near his person someone of our council in whom we have full confidence to perform the charge of intendant of justice. . . ." On other occasions the commission stressed the need to care for the army. Guillaume Bordeaux's commission as intendant of the army of the marshal de La Force in 1633 read: "It being necessary for the good of our service to allow someone of merit and the requisite sufficiency to have the charge and direction of our finances in the army. . . ." Likewise, that of Gaspard Du Gué, intendant of finance in the region and army of Provence in 1636, affirmed: "It being necessary for the subsistence, nourishment, and maintenance of our army that someone handle various large expenses. . . ." But the commission of Alexandre de Sève in 1637 said only: "Having at the same time judged it necessary to commit the administration of justice, police, and finances . . . [illegible words] to a person of our cognizance capable of acquitting himself worthily. . . ." At other times the rationale was much more specific. François Lasnier, for example, was named intendant of the army of Picardy in 1638 be-

2. Such common phrases were in all commissions; see, for example, the printed text of Claude Gobelin's commission, in Mousnier, *Lettres au Séguier*, 2:1047–49. I examined a wide variety in the war archives at Vincennes.

3. Quotations taken from the commisions of Bordeaux, Du Gué, and Voyer d'Argenson, A.G. A[1] 26, fol. 64; A[1] 12, fol. 131; A[1] 14, fol. 32, dated 1635, 3 February 1630, and 12 August 1632, respectively. There is some doubt whether Bordeaux's commission was actually issued, for on the back of the parchment copy in the war archives there is the notation "n'a été expedée," but Bordeaux did serve in this area during this period. Nevertheless, see the comments of Emile-Edmond Legrand-Girarde, *L'Arrière aux armées sous Louis XIII: Crusy de Marcillac, évêque de Mende, 1635–1638* (Paris, 1927), p. 23, who misquotes the back of Bordeaux's commission and ignores the fact that his commission was for the intendancy of the region around Champagne and in the army.

cause of reports reaching the government indicating that various officers and soldiers were leaving the army without permission and that
certain men along the frontier were having dealings with the enemy.[4]

In all commissions it was common to make reference to the character
of the man named to fill an office, and the commissions of the army intendants were no exception. The formula was standard and often trite.
For example, when Du Gué became intendant of finance in the army in
1629, his commission explained: "We do not know how to make a
more worthy choice nor better selection than of your person, for the
entire confidence that we have in your sufficiency, capacity, experience,
and vigilance." In certain cases, an especially valuable servant was
lauded more than others. Jean de Choisy, intendant of the province of
Champagne, received a supplementary commission in 1635, granting
him the charge of intendant of justice and police in the armies passing
through his province. The commission stressed his particular knowledge
of the area, his experience in past intendancies, and "your particular
affection and diligence for our service which you have fulfilled [in the
past] to our entire satisfaction." Indeed, Choisy had a long and meri-
torious career in the royal service, acting as intendant in the provinces
of Champagne, Roussillon, and Languedoc as well as various army intendancies. Later Choisy advanced to the position of chancellor to the
duke of Orléans and mediated the delicate negotiations between the
king and his obstreperous brother.[5]

The royal appointees were capable. Considering the importance of
an intendant's mission, it was natural to name councilors who had previous experience in the king's service. The career of René de Voyer
d'Argenson is an interesting example. He began his rise as an *avocat*,

4. Commission of Argenson cited in preceding note; commission of Bordeaux,
1 September 1633, A.G. A¹ 14, fol. 87; commission of Du Gué, dated 5 April, 1636,
A.G. A¹ 32, fol. 51; first draft of a commission for "sieur de St Jullien de Seve,"
dated March 1638, A.G. A¹ 42, fol. 81; commission of Lasnier, 16 August 1638,
A.G. A¹ 49, fol. 162. The date of Sève's commission seems to be incorrect; it
should be 1637 because it accompanies in the archives a project for a commission
for the intendancy of the army of Burgundy commanded by the duc de Longueville, dated 26 March 1637, A.G. A¹ 42, fol. 80. There is an identification problem
with this Sève. It must be Alexandre de Sève, later sieur de Chastignonville, although his father bore the title of sieur de Saint Jullien. The father, however, was
a banking figure who held the position of receiver general and paymaster of the
rentes of the clergy, and there is no evidence that he ever served as an intendant.
Alexandre was a young man at the time, he married during that same year, 1637,
and he probably took his father's title, perhaps as a marriage portion. See Mousnier, *Lettres au Séguier*, 1:121–28, for biographical details.
5. Commission of Du Gué, dated 1629, A.G. A¹ 13, fol. 149; commission of
Choisy, 28 April 1635, A.G. A¹ 26, fol. 49. For details of Choisy's life, see Prevost
and Roman d'Amat, *Dictionnaire de biographie française*, s.v. "Choisy, Jean II";
and Mousnier, *Lettres au Séguier*, 2:1194.

or barrister, who pleaded cases before the Parlement of Paris, then switched to the royal service as a councilor of state (1625) and a master of requests (1628). He assisted at the siege of La Rochelle (1627), before being commissioned to demolish the fortifications of Bergerac in 1629. In 1630 the crown named him to his first intendancy, that of the province of Dauphiné and adjacent regions. The next year he obtained an appointment as *procureur général* of the Arsenal, charged with stopping the counterfeiting of money. In 1632 he served as intendant of justice, police, and finance attached to the prince of Condé in Limousin, Marche, and Auvergne; the following year he was intendant of the provinces of Saintonge and Poitou.

Then Argenson supervised the demolition of various rebel châteaus in Auvergne and Bourbonnais, obtaining the title of intendant of Auvergne in 1634. In 1635 the government named him intendant of the army commanded by the king in person, after which he served as intendant in the duke de La Force's army (1636) and in 1637–40 as intendant of the army in Italy. Next the crown shifted him to Catalonia, where he remained as army intendant, but he assumed extra duties as one of the commissioners named to negotiate the cession of that country to France (1641). Returning to France, Argenson obtained a new appointment as intendant of the provinces of Poitou, Saintonge, and Angoumois in 1644. Two years later he traveled to Italy for negotiations with the pope, the grand duke of Tuscany, and other Italian princes. Shortly thereafter he became the intendant of the Franco-Italian army under the command of Prince Thomas of Savoy. In 1647 Argenson served as special commissioner to the assembly of the estates of Languedoc, and in 1649 for the pacification of Guyenne. His last mission was as ambassador to Venice, where he died shortly after his arrival in 1651.[6]

Argenson's career is typical because of the mixture of provincial and military intendancies. For this reason, the key paragraph in a commission was that which delineated an intendant's authority: "For these causes and other considerations that move us, we have commissioned, ordered, and deputized you; we do commission, order, and deputize you by this present letter, signed with our hand, intendant of ———." An intendant might be named, as was Claude Gobelin in 1636, "intendant of justice, police, and finance in our army," or he might be named for only part of these duties, as was Guillaume Bordeaux in 1633, "intendant of our finances in our armies." Sometimes if an in-

6. Mousnier, *Lettres au Séguier*, 1:89–90. See also Alfred Barbier, *Notice biographique sur René de Voyer d'Argenson, intendant d'armée du Poitou, ambassadeur à Venise (1596–1651)* (Poitiers, 1885), pp. 4–32.

tendant was named only intendant of finance, the government would add other duties to his title, as happened to Gaspard Du Gué in 1629, when he was named "intendant of finance, supplies, munitions, and magazines" in the army of the marshal de La Force. If the crown commissioned a man to be only an intendant of justice, not an intendant of finance, his title might read like that of Sublet de Noyers in 1633, "intendant of justice and police of our army."

Provincial intendants were also intendants of justice, police, and finance, but no military duties were mentioned in the title, only the geographical jurisdiction. For example, Faucon de Ris in 1643 was "intendant of our province of Lyonnais, Forez, and Beaujolais and the generality of Lyon." Often the intendancy of an army and a province would be combined, and the commission might read like that of Sieur Ollier in 1640, "intendant of justice, police, and finance in our army and province of Guyenne," or like that of François Cazet de Vauxtorte in 1640, "intendant of justice, police, and finance in our land of Provence and adjacent territories, and in our army that we are assembling in our said province." [7]

When Vauxtorte served as intendant of the occupied German territories in 1645, his commission titled him "intendant of justice, police, and finance in the bishopric of Spires, archbishopric of Mainz, marquisate of Baden, Lower Palatinate, and other lands and places held by our armies in those quarters," and ordered him to undertake the "direction and intendance of justice, police, and finance in regard to the troops and in all things relating to our service." [8]

On occasions when a man was already an intendant of a province and he received the intendancy of an army operating in that region, it would not be necessary for the government to send him a regular commission, but only a supplementary one similar to that of Jean de Choisy in Champagne in 1635, "to perform and exercise the charge of intendant of justice and police in our armies." Sometimes an official might be named to substitute for an ill or absent intendant; he would

7. Commission of Gobelin, 10 August 1636; Bordeaux, 1 September 1633; Du Gué, 1629; Des Noyers *fils*, 1633 (I assume this is Sublet de Noyers, who was army intendant at this time. See Charles Schmidt, "Le Rôle et les attributions d'un 'intendant des finances' aux armées, Sublet de Noyers, de 1632 à 1636," *Revue d'histoire moderne et contemporaine* 2 [1900–1901]: 156–75); Ollier, 8 May 1640; Vauxtorte, 12 May 1640; A.G. A¹ 32, fol. 142; A¹ 14, fol. 87; A¹ 13, fol. 149; A¹ 14, fol. 133; A¹ 62, fols. 136, 139. The commission of Faucon de Ris, 18 May 1643, is in Mousnier, *Lettres au Séguier*, 2:1059–62.

8. Commission of Vauxtorte, 21 January 1645, in Mousnier, *Lettres au Séguier*, 2: 1071–75. There is a duplicate of this in the war archives, misdated 1633, A.G. A¹ 14, fol. 110.

not receive a normal commission, but only permission to exercise the charge in the absence of the intendant.[9]

INTENDANCE OF JUSTICE

After giving an intendant's full title, the commission enumerated his duties. The section relating to his responsibility in matters of justice preceded all others. Every commission specified that the army intendant should "reside close to the person" of the general "to assist at the councils which will be held there and to give your good advice." These *conseils de guerre* were of two main types. First, they were assemblies of the major officers in the army, who met together to deliberate on an important or difficult question, such as military strategy. They also were meetings at which military justice was meted out to offenders. Here the intendant acted as the chief judicial figure.

The above clause authorized the intendant to give technical advice in the first of the two circumstances. Since the intendant was expected to have an exact inventory of all the army's supplies, he could respond to a general's questioning at such a council. The intendant provided the necessary information about the army's condition, which could spell the difference between success and failure. It was for this reason that the army intendant had a consultive voice in the general's councils. One army intendant, Robert Arnauld d'Andilly, recounted an instance of technical counseling in his memoirs. Once when Arnauld was too sick to attend a council meeting, a Colonel Hebron proposed that five thousand cavalrymen, each carrying a sack of grain, transport these supplies to the Rhine, where they could be sent by boat to relieve the siege of Mannheim. Informed of the council's decision to accept this scheme, Arnauld boldly declared it to be folly, for the distance between them and Mannheim was great and the enemy lay between. Sure that the opposing forces would hear of the French approach, Arnauld speculated on the consequence of the cavalry's being surprised with their sacks. They would immediately cast them aside in order to defend themselves, thus incurring the loss of the provisions. When he told the officers, the marquis de Feuquières and the cardinal de La Valette, his opinion, they took his advice and abandoned the plan.[10]

9. Commission of Choisy, 28 April 1635, A.G. A¹ 26, fol. 49; commission of Dorée to exercise the charge of intendant of finance in the army of Catalonia during Argenson's absence, 12 December 1641, A.G. A¹ 67, fol. 234.

10. Pierre-Claude de Guignard, *L'Ecole de Mars ou Mémoires instructifs sur toutes les parties qui composent le corps militaire en France, avec leurs origines, et les differentes maneuvres ausquelles elles sont employées* (Paris, 1725), 2:339; Robert

Participation in a general's councils was important not only for advising the commander, but also because it enabled the government to place a trusted agent beside a powerful general. Sitting in the military councils, the intendant always knew exactly what was happening, and having the minister's ear, he did not hesitate to inform the government of potential treason. This was especially vital in the early decades of the seventeenth century, when the crown tried to curb the power of the "great ones" who did not scruple to cabal against the king.

There was an excellent example of this in the summer of 1632, when the state named René de Voyer d'Argenson as intendant "close to our dear cousin," the prince of Condé. Argenson's commission included the standard phrase about the importance of having someone near Condé's person to assist him, but one must not be deceived by the polite wording. In reality, Condé was in royal disfavor. The "Day of Dupes" had occurred the previous year, the queen mother was exiled, and Monsieur, the king's brother, fled the country. Although suspected of treason, Condé reconciled himself to the king, but he had offended Richelieu when the cardinal desired that the prince take the odious job of chief judge of the tribunal that was to try the popular Marshal de Marillac for treason. Condé declined this dubious honor on the grounds of ill health, but Richelieu was not deceived by his excuses. At that very moment Monsieur openly revolted against the king and entered Languedoc, where he was welcomed warmly by Henry II, duke of Montmorency and governor of the province. Condé and his friends feared that he would be named to head the army against Montmorency, his own brother-in-law, but Richelieu did not dare put Condé to the test; instead the cardinal occupied the prince's time by ordering him to chase outlaws in Limousin, Marche, and Auvergne. As extra insurance for Condé's fidelity, Richelieu named Argenson as army intendant to accompany him.[11]

The commission of an army intendant spelled out other duties besides assisting the general in his councils. For example, he was to "see that justice is sincerely administered in our army and that crimes and thievery are chastised and punished according to the rigor of our ordinances, which we wish to be exactly kept and observed." To this, the commission sometimes added the injunction to see that the troops

Arnauld d'Andilly, *Mémoires de Messire Robert Arnauld d'Andilly*, ed. Petitot, Collection des mémoires relatifs à l'histoire de France, 2d ser. (Paris, 1824), 34:64–66. Cf. also the brief comment of André-Eugène Navereau, *Le Logement et les ustensiles des gens de guerre de 1439 à 1789* (Poitiers, 1924), p. 47

11. Commission of René de Voyer d'Argenson, 12 August 1632, A.G. A¹ 14, fol. 32. Henri d'Orléans, le duc d'Aumale, *Histoire des princes de Condé pendant les XVIᵉ et XVIIᵉ siècles* (Paris, 1863–96), 3:232–42.

"live in good discipline and police." In order to fulfill this duty, the crown urged the intendant to "hear and receive complaints and grievances that will be made to you by our subjects or others of wrongs, exactions, and violences that will be done to them and render them good and quick justice." [12]

One might ask how military justice was carried out in practice. Although an intendant's commission ordered him to "inform himself of all enterprises, practices, and attempts that will be made, or have been made against our service," no matter what the rank of the culprit, it seems that the intendant usually confined himself to examining evidence compiled by others. The key officials in criminal cases were the provost general of the army and his subordinate provosts, the *prévôts des bands*, who combined the functions of policemen, prosecuting attorney, and, in some senses, jury. With bands of archers, the provosts patrolled the army looking for infractions of military procedure. "See and visit as many times as you feel necessary all the quarters and places of the armies," declared the commission of a provost general in the army of Italy in 1636; "inform yourself of contraventions to our *ordonnances* and *règlements* which will occur in our said armies." Once the archers had captured a suspect, a provost and his clerk conducted the preliminary investigation, drawing up a document cataloging the charges. This indictment became the basis of the trial and was delivered to the intendant, who presided as chief judge at a *conseil de guerre*. Secondary judges assisted him in reaching a verdict.

According to Bassompierre's memoir, these judges consisted of the provost general, his subordinate provosts—if the intendant judged them suitable—or officers in the camp with a legal background. Lacking these, the intendant was supposed to draw up an order, signed by the general, requiring judges from the nearest locality to assist at the trial. The trial was conducted according to the codes of military justice, and the court's decision was final. "We validate," an intendant's commission decreed, "the judgments that will be thus rendered by you as if they had emanated from our sovereign courts, and we desire that they be executed notwithstanding any opposition or appeals whatsoever." [13]

Many matters of military justice were handled without trials. The provost had authority by his own commission to punish "on the field"

---

12. Common phrases in most commissions; cf., for example, that of Gobelin, in Mousnier, *Lettres au Séguier*, 2:1047–49.

13. Commission of Sieur Le Grain, *lieutenant criminel* of Paris, named 24 February 1636 as provost general of the armies of Italy, A.G. A¹ 32, fol. 29. The details of the criminal procedure are from Bassompierre's memoir in B.N. *cinq-cents Colbert* 499, fol. 99. Those with a legal background were *graduez*: people who had studied at a *faculté de droit* and obtained a rank such as *bachelier* or *licencié*.

those caught in flagrante delicto. D'Estampes, intendant of the army in Italy in 1630, wrote to the government that the provosts who watched the mountain regions bordering France and Savoy needed a royal writ to permit them to hang deserters "without any form or figure of a trial," remarking that he had already given such permission in the army in order that the provost's men could search for soldiers absent without leave and hang them along the main routes. Attached to the executed man would be a notice containing the grim admonition "soldier absent without leave." In another instance, Richelieu wrote to Du Gué, another army intendant, to extract the sum of 24,000 *livres* entrusted to an official who had not performed his task. "Get back this sum," the cardinal ordered, "and without any formality of a trial, imprison him."[14]

Even if there was a trial, there was often little doubt as to the verdict. On 4 September 1640, Richelieu wrote to Nicolas Bretel de Grémonville, intendant of the army of the marshals de Chaulnes and de Chastillon, that approximately one hundred deserters, including some officers, had been arrested at Amiens and were being sent back to be judged in a *conseil de guerre*. The cardinal ordered Grémonville to see that "you lose no time in the said judgment, and also that you take care to accomplish it, as a thing absolutely necessary for His Majesty's service." He warned that an example must be made of the officers "because without that, it is impossible to maintain armies." As soon as Grémonville completed the task, the cardinal asked him to report back.[15]

In such actions the army intendant was often indistinguishable from the "hanging judges" that roamed the recently pacified provinces, administering summary justice. A prime example is Isaac Laffemas, the notorious intendant known to history as "Richelieu's executioner." On 6 February 1633, Laffemas received a commission as "intendant of justice, police, and finance as much in our said armies as in the towns of the province of Champagne [and in] Metz, Toul, Verdun, and other places of our obedience and protection and everywhere else our armies can extend." The royal armies had recently subdued this region, but the area refused to stay pacified. "In place of renouncing their cabals, new

14. See the commission of Le Grain cited in the preceding note. Letter of J. d'Estampes, 3 September 1630, from the camp at Veiglane, A.A.E. *Mémoires et documents FRANCE* 796, fol. 250. This letter is also in Hanotaux, *Origines des intendants*, pp. 336–39. Richelieu to Gaspard Du Gué, 8 April 1630, in Denis-Louis-Martial Avenel, ed., *Lettres, instructions diplomatiques et papiers d'état du cardinal de Richelieu*, Collection de documents inédits sur l'histoire de France (Paris, 1853–77), 3:614–15. See also "Instruction pour le sieur de Lisle," 9 April 1630, confirming Du Gué's orders, in Avenel, *Lettres de Richelieu*, 3:618–21.

15. Richelieu to Grémonville, 4 September 1640, in Avenel, *Lettres de Richelieu*, 6:720–21.

ones are formed every day," the intendant's commission claimed, threatening the frontier and seducing loyal servants from royal obedience. Laffemas's mission was twofold—he was to try the rebels for their crimes and to see that the troops lived in discipline, not oppressing the king's subjects. The intendant fulfilled his task zealously. On 6 March 1633 he wrote from Troyes: "I have issued a warrant for the arrest of thirty-four gentlemen and others who have risen against the king and I am ready to order the same against eight others. . . . They are absent for the most part and must be judged by default." Two weeks later he issued charges of rebellion against other gentlemen who had borne arms against the king at Castelnaudari. They were to be tried the following Wednesday, and Laffemas promised: "We can arrange the different punishments so that all go to death." These trials continued throughout the summer months and far into the autumn; even as late as November, such unfortunates as the chevalier de Jars were tried and sentenced to death.[16]

Laffemas did not spend all of his time trying past offenders. He also sought to root out possible treason, particularly among the religious orders angry over the government's alliance with Protestant princes. In November he intercepted treasonable correspondence between some monks and investigated fathers of the Recollect and Minim orders traveling between Brussels, Besançon, and Italy. The following month he seized a letter from a Father Mahaut addressed to Dame de Pontcarré at Port Royal that hinted at treason, so he recommended that the woman's rooms be secretly searched for further evidence of her guilt. During that same month of December a monk from Saint-Menge was captured and tortured. Despite his ordeal, the victim refused to confess, but Laffemas vowed that he would "put him behind four walls to finish out his days." [17]

16. Laffemas's commission is in A.G. A¹ 14, fol. 44, and in Hanotaux, *Origines des intendants*, pp. 322–26, although the two differ slightly in wording. Laffemas to Séguier, 6 and 20 March 1633, in Mousnier, *Lettres au Séguier*, 1:197–200. For a popularized biography of Laffemas, see Georges Mongrédien, *Le Bourreau du cardinal de Richelieu, Isaac de Laffemas* (Paris, 1929). Laffemas did not like his reputation as "Richelieu's executioner." For example, he wrote to Chancellor Séguier, 27 September 1636, "et voudrois qu'il pleust à Sa Majesté terminer là mes emplois criminels et me donner moyen de la servir en autre chose. J'aurois bien de l'obligation à vostre bonté, de m'avoir procuré ce repos là, pour ne plus passer pour un homme de sang, en faisant la justice, qui est en ce temps odieuse à beaucoup de gens qui ne sont point touchez de l'intérêt public" (in the appendix of Monmerqué and Paris, eds., *Les Historiettes de Tallemant des Réaux*, 3d ed. [Paris, 1854–60], 5:535–36). The *arrêt* of Laffemas condemning the chevalier de Jars is found in this same *Historiettes*, 5:512–13n1. Jars was not executed, but imprisoned in the Bastille.

17. Laffemas to Séguier, 8 and 11 November and 7 December 1633, in Monmerqué, *Historiettes*, 5:509–12, 513–16, 527–30; Laffemas to Séguier, 11 December 1633, in Mousnier, *Lettres au Séguier*, 1:217–18.

Laffemas's mission illustrates the curious intermixture of the military and civilian nature of an intendant. The above examples have examined his work among the general populace, punishing rebels and searching out treason; at the same time he also fulfilled his military functions. There is an excellent illustration of this in 1636, when Laffemas, at Senlis, reported in one breath the condemnation of a baron de Senac, and in the next the punishment meted out to canteen managers who bought stolen goods from the soldiers. They were banished from the army, condemned to the lash, and branded with the fleur-de-lis before being locked in a pillory.[18]

Sometimes the intendants in the army could not handle the double burden of trying suspects and caring for the army. In 1636, the terrible year of Corbie, the Germans invaded Burgundy while the Spanish advanced into Picardy and Guyenne. The situation on the eastern front reduced the French armies to desperation: they were forced to provision themselves at the edge of the battlefield, the unpaid armies pillaged the countryside, the peasants fled, and disloyal towns surrendered to the Spanish without a blow. Gobelin, intendant of the army of Picardy, and Argenson, intendant in the army of the duke de La Force, could not handle the problem alone, and the government sent in reinforcements. Masters of requests Aubray, Choisy, and Orgeval arrived in the armies to act as temporary intendants, handling matters of supply and punishing those guilty of intelligence with the enemy. In a letter from the camp of Péronne, dated 23 September 1636, Argenson wrote to the chancellor on behalf of Choisy, Gobelin, and Orgeval, asking if they should proceed with the trial of the lieutenant general suspected of treason at Roye or if they should send the case to the chancellor for his opinion. Meanwhile, Orgeval continued to search out traitors, uncovering a guilty notary in addition to the lieutenant general and a *procureur du roi*, justifying Argenson's suspicions that a large number of people were involved in surrendering the town of Roye to the enemy.[19]

Throughout Richelieu's lifetime, intendants continued to investigate charges of disloyalty to the king; whether they did this in their capacity as intendants of an army or as intendants of a province was a matter of little difference. Nothing illustrates this interchangeability

18. Laffemas to Séguier, 26 September 1636, in Monmerqué, *Historiettes*, 5:533–35.

19. Boris Porchnev, *Les Soulèvements populaires en France de 1623 à 1648* (Paris, 1963), pp. 52–54. Notice of the commissions of Claude Gobelin and René de Voyer d'Argenson are found in A.G. A¹ 32, fols. 142, 178. Aubray to Séguier "au camp de Lussi," 21 September 1636, in Mousnier, *Lettres au Séguier*, 1:309–10; it bears the notation "Daubrav maître des requêtes faisant la charge d'intendant dans l'armée." Argenson to Séguier, 23 September 1636, in Mousnier, *Lettres au Séguier*, 1:310–11.

better than an incident in the latter part of Louis XIII's reign. In 1641, Gobelin, intendant of the army of Picardy, began an inquiry into charges that the officers and soldiers of the garrison of Lillers had abandoned their post instead of defending it courageously against the enemy. But events called Gobelin elsewhere, and he was forced to absent himself from the inquiry; in his place the government instructed Louis Le Maistre, sieur de Bellejamme, intendant of justice in the province of Picardy, to conduct the trial. Unfortunately the documents that reveal the denouement of this tale are missing, but it is clear that there was still not a great distinction between the provincial and army intendants during the 1630s.[20]

INTENDANCE OF FINANCES

In addition to his judicial duties, the intendant of the army had certain responsibilities as intendant of finance: his commission ordered him to "have an eye to the handling and distribution of the moneys which are destined for the pay and maintenance of our men of war." Some commissions enumerated another half-dozen expenses assigned to an intendant's care: light cavalry, infantry, artillery, victuals, munitions, reparations, fortifications, and unforeseen expenses.

Second, the intendant had the task of auditing the army's accounts. "See, verify, and regulate," read an intendant's commission, "the bills, statements, and orders [for payment] which will be sent by our lieutenants general to the treasurers general, whether of the ordinary or of the extraordinary funds, for war." To this was added the chore of verifying the various extracts, rolls, and registers of payments and reviews made to the troops by the commissioners and controllers of war.[21]

Such a responsibility must have been a burdensome task to the intendant, with the compensating excitement of scandal a rare phenomenon, for few letters speak of such problems. Yet some cases have come to the historian's attention. Charles Turquant, intendant of the army in Brittany at the end of the sixteenth century, investigated charges of embezzlement against a man who was supposed to pay the troops on the basis of their reviews; the criminal was found guilty of "beaucoup

20. Letters patent dated 28 August 1641 ordering Bellejamme to go to Montreuil and proceed with the trial because of Gobelin's departure, A.G. A¹ 65, fol. 419.

21. These phrases are from the commission of Du Gué, intendant of the army in Burgundy, 1629, and Amelot de Beaulieu, intendant of the army of Guyenne, 10 February 1639, A.G. A¹ 13, fol. 149, and A¹ 56, fol. 46, but some variation of them is found in most commissions. See also Bassompierre's memoir, B.N. *cinqcents Colbert* 499, fol. 99.

de faussetés" and hung. Later Turquant instituted proceedings against Paul Robichon, former treasurer provincial of the extraordinary funds for war, also suspected of embezzlement.[22]

Sometimes a commission of an army intendant listed responsibilities for supplies, especially if he was only an intendant of finance, for it was assumed that he would have the additional time needed to care for the victuals, munitions, and magazines. Consequently these items were added to his commission. Gaspard Du Gué is an example. His commissions as intendant of finances, supplies, munitions, and magazines in the army of Italy (1630) and as intendant of finances in the lands and army of Provence (1636) give detailed instructions regarding supplies.[23] In 1630 Du Gué was in charge of all the supplies needed for the king's campaign in Savoy. Not only was he responsible for the care and distribution of these provisions, but he also had the authority to purchase grain, oats, clothing, implements of war, and all other things necessary for maintenance of the armies. His commission granted him authorization to borrow whatever money he needed, permitting him to reimburse the principle plus interest from the general receipts of Lyon, an easy maneuver because Du Gué was also the treasurer of France in the bureau of finances at Lyon. Meanwhile, other royal officials levied contributions in money and in rations on the districts of Bresse, Bugey, Valentinois, Gex, Lyonnais, Forez, and Beaujolais. The intendant's task was to oversee the collection of these payments and to keep the accounts, sending them to the royal council for examination and approval.

Once the supplies were obtained, they had to be cared for, and the intendant had the authority to make any contract or agreement that he felt necessary to manage them at the most reasonable price. To assist Du Gué in the care and distribution of the supplies, he could appoint such assistants (*commis*) as he needed, retaining or dismissing them as he felt appropriate. Next, he was concerned with their transportation to the theater of war. For this task, the commission recommended that Du Gué remain at Lyon and arrange the departure from there, overseeing the loading of supplies, including cannon and munitions, on mule- and horse-drawn carts. The intendant could requisition

22. Séverin Canal, *Les Origines de l'intendance de Bretagne: Essai sur les relations de la Bretagne avec le pouvoir central* (Paris, 1911), p. 39. The war department had two types of funds, those of the ordinary and those of the extraordinary, hence the unusual-sounding title of treasurer provincial of the extraordinary.
23. The commissions of Du Gué are in A.G. A¹ 12, fol. 131, and A¹ 32, fol. 51.

the animals and their carts wherever he found them, conscripting their owners to serve as wagon masters in the army for as long a period as he felt necessary. Finally, the commission warned Du Gué to watch carefully the distribution of the supplies, "preventing that there be committed any abuse or malversation to the prejudice of our service, or the oppression of our people."[24]

Although Du Gué's commission of 1630 gave more detailed instructions than most, his commission as intendant of the army and lands of Provence in 1636 succinctly summarizes the typical intendant's responsibility for supplies: he was to see to the conservation and distribution of provisions, making such contracts and appointing such assistants as he needed; he was to submit a financial accounting of the various expenses involved in the supply; and he was to see that no pay or victualing was made except upon the basis of exact reviews of troops held in his presence and listed on rolls that he personally inspected.[25]

Such duties agree with Bassompierre's memoir, which added that the intendant had the responsibility for the magazines and storehouses of grain, flour, and supplies, together with their convoy to the army. Again, the memoir warned, he was to oversee their distribution only on the basis of reviews made by the *commissaires des guerres* and in accordance with the military ordinances of the *maréchaux de camp*. These latter documents specified the quantity of supplies to be furnished to each regiment and company of infantry. By watching the number of troops and the rate of bread distribution, the memoir suggested, an intendant could estimate ahead so that no dearth occurred.[26]

How well did Du Gué carry out such instructions during the campaign of 1630? A few of Richelieu's letters written during the campaign offer revealing insights. The cardinal stressed the importance of supply, believing that by its lack an army was lost. The king's presence, too, meant that Richelieu insisted that everything go smoothly, for the minister's power was not so entrenched as to withstand the resentment of a monarch who felt himself humiliated. Consequently Richelieu personally assumed much of the responsibility for supply. Before the campaign, he wrote to Marshal de La Force, who was in command of another army in the Alps: "I will bring to bear such care as to see that everything necessary is done for the subsistence of the army of His Majesty and I can assure you that nothing desirable will be lacking

24. This and the above paragraphs are from Du Gué's commission in 1630, A.G. A¹ 12, fol. 131.

25. Du Gué's commission as intendant of the army and *pays* of Provence, 1636, A.G. A¹ 32, fol. 51.

26. Bassompierre's memoir, B.N. *cinq-cents Colbert* 499, fol. 99.

for this end." Richelieu kept his word. Upon his arrival in the army, he distributed twenty thousand suits of clothing that had been ordered by the king.[27]

Du Gué's role, therefore, was subordinate to the cardinal's in the matter of supply. In early March, Richelieu wrote the king that Du Gué had left for Lyon with orders to prepare fifty carts and raise eight hundred horses, but that it would be necessary for the surintendant of finance to press the intendant and send money for payment. In April, Richelieu wrote to Du Gué asking him to speed the departure of several gun carriages, six pieces of cannon, and the thirty to forty horses needed to haul them to Grenoble for transport to the army; mules were also necessary. The cardinal instructed the intendant to obtain the animals, brand them, put them on the king's payroll, and then send them on to Grenoble, where the intendant would order a *commissaire des guerres* to conduct them to the army. But the French needed more than mules, for transport meant little without provisions; grain had to be collected from the Bresse region in as large a quantity as possible, or else the army would not have enough. "Do not lose any time," the cardinal warned the intendant, and "see diligently to all that I have written you and do the things you think necessary without awaiting my response." Meanwhile, Richelieu assured Marshal de La Force that Du Gué had already gathered some hundred twenty horses to draw cannon and could obtain more, adding, "S^r Du Gué will furnish the supplies. It has been so long since I have written to him on this, that I do not doubt that he has a munitioneer and the grain." [28]

Du Gué did sign an agreement with a munitioneer to furnish bread for the army, but this did not satisfy the anxious cardinal. In a memoir on the conquest of Savoy, Richelieu recounted that on his arrival in Grenoble, two days before that of the king, he found no supplies ready to follow the army. He therefore made his own arrangements, buying six thousand loads of grain on his own credit; but the *maréchal de camp*, Du Hallier, and intendant Du Gué said there was no need for this, because they had already made a contract with a munitioneer who promised to furnish the victuals while the army was in Savoy. Richelieu sarcastically commented: "The little experience that I have had in such affairs told me that such men often undertake what they cannot

27. For the importance that Richelieu placed upon the problem of supply, see Legrand-Girarde, *L'Arrière aux armées*, pp. 2–3. Richelieu to La Force, 10 December 1629, in Avenel, *Lettres de Richelieu*, 3:481.

28. Richelieu to the king, 9–10 March 1630; Richelieu to Du Gué, 8 April 1630; "Instruction pour le sieur de Lisle"; Richelieu to the marshal de La Force, "vers le 10 d'avril 1630"; all in Avenel, *Lettres de Richelieu*, 3:573–80, 614–15, 618–21, 624–25.

fulfill and do not recognize the difficulty of the enterprise . . . but promise [instead] what they do not know how to effect." The cardinal had spoken with wisdom, since the king was forced to wait five or six days at Barrault for the supplies.[29]

Increasingly, intendants occupied themselves with the problem of the army's subsistence as time passed, and less with matters of justice. It became the intendant's duty to watch the munitioneer, giving him all the assistance needed to obtain provisions, yet at the same time acting as a control against possible wrongdoing. In the disastrous year of Corbie (1636), army intendant Choisy's efforts illuminate this. He reported that the munitioneer was busy at work, for he had dispatched his assistants into the field to subcontract for the army's bread, and Choisy had written to Governor Monbazon to assist these men in any way possible, particularly by permitting them to requisition the inhabitants' provisions in return for payment. As a check against corrupt practices, Choisy sent the army's *contrôleur des vivres* to supervise the munitioneer's agents. In the meantime, he wrote letters to the mayors and city fathers of the district urging them to exchange flour for unground grain, while he searched for additional mills to speed the production of flour.[30]

Grain was difficult to obtain in the troubled year of 1636, and in many cases the intendant had to go out and look for it himself. Choisy traveled to Noyon to find grain and then shipped it to Saint-Quentin and Ham, where additional carts transported it to the army, while Orgeval rode to Amiens searching for more cereal to make into flour. Aubray, another master of requests, summarized all the difficulties involved in supplying the army in a letter he wrote 21 September 1636 from the camp of Lussi. He reported that enough supplies were on hand around the river Oise, but there was only enough food near the Somme River to feed about one-third of the troops. He thought that Amiens was the most likely place to obtain the needed supplies, but the enemy garrisons at Corbie and Bray prevented the French from expecting much help from that source. Only Péronne and Ham could definitely be relied upon, and they could give no more than ten thousand loaves of bread. Saint-Quentin lay only five leagues from Péronne, but the distance actually traveled by the supply carts made this route exceedingly difficult because the road followed the bends in the river.

Noyon was the only feasible alternative. Bread could be obtained

29. Details from a "minute" in the hand of Charpentier, 7 July 1630, concerning the expedition to Savoy, in Avenel, *Lettres de Richelieu*, 3:738–42.

30. Jean de Choisy to Séguier, 1 September 1636, in Mousnier, *Lettres au Séguier*, 1:299–300.

there, but the army would need a hundred additional carts for transport, since each cart had a capacity of only six to seven hundred loaves and the route from Noyon to Péronne took five days. The hundred twenty carts that the army already possessed hardly sufficed for half the army's needs, even if they worked the horses without rest—a thing manifestly impossible. Aubray hoped to find extra carts at Noyon, for he had received a report from an archer of the provostship that there were seventy carts available. He promised to get the carts despite the expense; "without them the army will be forced to disband." [31]

From reading such reports, it is obvious that by the 1630s the central government for the most part had capable agents in its armies, men who had mastered the complexities of the problems and found solutions. The importance of these army intendants was overwhelmingly apparent during the crisis years of 1635–37, when France's territorial integrity was threatened during the Thirty Years' War. The intendants, working behind the lines, organized a massive system of supply—the vital flour for bread, the cannon, and the powder upon which France's defense depended.

Part of this credit, certainly, belongs to Richelieu. Recognizing the importance of an efficient organization of supply, he personally supervised much of the intendants' efforts, prodding or pressuring as necessary. Those intendants who were lax in their duties received blistering letters from the cardinal. For example, Guillaume Bordeaux, intendant of finances, supplies, and magazines in the army of Champagne in 1635, was supposedly hard at work in Champagne carrying grain to the army, but Richelieu received reports from his agents that although the intendant should have arrived in the region fifteen days earlier, in reality he had left Paris only three days before. The cardinal warned: "I do not know if I have enough credit with the king to turn His Majesty from the just resentment that he must have of your laziness and lack of affection for his service. . . . and, in fact, you cannot deny that a lackadaisical negligence such as yours is as criminal as open malice. It is up to you to work diligently so that the armies of the king lack nothing." [32]

IMPLIED FUNCTIONS OF AN INTENDANT

The commissions enumerated the powers granted to an army intendant, both explicit and implicit. Not only did an intendant have to

---

31. Chavigny to Séguier, 23 September 1636, and Aubray to Séguier, 21 September 1636, in Mousnier, *Lettres au Séguier*, 1:352–53, 309–10.
32. Richelieu to Bordeaux, "1635, juillet," A.A.E. *FRANCE* 815, fol. 54.

fulfill certain functions, but he obtained wide authority for this purpose. The closing section of one commission remarked: "Do this and generally everything else that you recognize and will judge necessary for the execution of your present commission." Another commission put it this way: "And generally proceed to the execution of everything in the above circumstances and dependencies that you see needs to be done in our service for the well-being and comfort of our subjects and the observation of our ordinances. To do this, we give you power, permission, authority, and special command by these present letters." These implied powers were the real basis for an intendant's authority, and the intendants soon assumed functions well beyond the immediate realm of their commissions. As agents of the crown, they gained control over the older institution of the *commissaires des guerres*. Intendants also had charge of military hospitals, the care of the wounded, and special projects such as fortifications, although none of these was mentioned in their commissions.[33]

The intendant's commission usually had a clause allowing him to name *commis*, or assistants, but nowhere was there any specific delegation of authority over the *commissaire des guerres*. Yet these officials became subordinate to the intendants. There is no study that illuminates this process, but fragmentary evidence gives certain glimpses without telling the whole story. Certainly by the 1630s the *commissaires* acted under intendants' orders, for Arnauld d'Andilly recounted in his memoirs that the military commissioners inspected the troops on his instructions. But when Le Tellier was army intendant in Italy in 1640, a commissioner named Talon stationed at Turin had orders to pay the troops in his own reviews, without going through the intendant as an intermediary. Talon corresponded directly with the secretary of state for war and the surintendant of finance, and did not seem to be dependent upon an intendant. Between these two extremes, little is known of the relationship between the intendants and the military commissioners in the 1630s, but after Le Tellier assumed the secretaryship of war, there is abundant evidence of the intendants' domination of the commissioners.[34]

The implications of this dominance are enormous: supported by royal authority, the intendants constructed a chain of command from the secretary of state for war at the top to the commissioners at the lowest level. The intendants were the intermediary links, vital links of

33. Commissions of Gaspard Du Gué, 3 February 1630, and François Lasnier, 16 August 1638, A.G. A¹ 12, fol. 131, and A¹ 49, fol. 162.

34. Arnauld d'Andilly, *Mémoires*, 34:47–49. Narcisse-Léonard Caron, *Michel Le Tellier: Son Administration comme intendant d'armée en Piémont, 1640–1643* (Paris and Nantes, 1880), p. xlvii.

control, over scores of local agents who handled most of the details of everyday administration. The team effort of *commissaires des guerres* and intendants, their esprit de corps, meant that the crown had at last found a team of administrators on whom it could depend, and it was through the efforts and obedience of the commissioners that the intendants made themselves felt throughout every aspect of army life.

Army intendants were responsible for the establishment of ambulatory hospitals and the care of the wounded. The memoir attributed to Bassompierre remarked: "The intendant . . . must often send [assistants] and even go in person to see how the wounded and the sick are assisted, treated, and bandaged; which charge, in addition to fulfilling his duty and being charitable, brings him great goodwill and reputation in the army." Consequently it is not surprising that there exists a letter from Argenson, intendant in the army of Marshal de La Force in the troubled days of 1636, expressing concern that the weather was so frightful that soldiers who stood guard duty fell ill and a large number of sick filled the hospital at Glizay. Argenson inquired if another hospital could be established, "for we have no remedies nor nourishment to give, except their bread." It was not, however, until the administration of Le Tellier and his son Louvois that establishing proper medical facilities for France's soldiers constituted one of the major duties of the army intendants.[35]

The situation is similar with regard to work on fortifications along the frontier. True, some intendants like Sublet de Noyers received powers in addition to their regular commissions. Sublet was a "commissioner deputized by His Majesty for the fortifications and provisionment of the places of the province of Picardy." Sublet's task, however, was merely supervisory. The real job of engineering was left to a professional, Argencour, yet Sublet did make inspection trips to see that the garrisons were well fortified, the officers doing their duty, and that they were abundantly supplied. Similarly, Charles Machaut, intendant of the army of Burgundy in 1636, reported the sad condition of the fortifications at Dijon: the place was in a sorry state, work already begun had ceased entirely due to lack of money, and the citizens were unwilling to contribute—the privileged families protested rather than paid. Moreover, Argenson's instructions as intendant of the army and land of Catalonia in 1641 ordered him to fortify the promontory of Mont Julich, which commanded the heights around Barcelona, and to establish a French garrison there "so that it will not be in the power of the king of Spain to take it." But these were unusual cases.

35. Bassompierre's memoir, B.N. *cinq-cents Colbert* 499, fol. 99. Argenson to Séguier, 1 November 1636, in Mousnier, *Lettres au Séguier*, 1:322–23.

Not until the reign of Louis XIV did the government really become interested in fortifications along its frontiers, and then it would be the frontier intendants who assisted Vauban in this great work.[36]

Regardless of the multitude of duties, the intendant's most important function was to keep the government informed of everything that occurred in the army. Bassompierre's memoir underscores this necessity: "The means to make one important at court is to describe frequently to the secretary of state, the surintendant of finances, and those who govern, the state of the army, of a siege if there is one, of the force of the troops, of the quantity of officers, of the needs of the army, of the hospital, and of supplies. Let them also know what can be learned of the enemy forces, of their designs and projects, and generally what he [the intendant] judges apropos to tell them." This was exactly what the intendants, the loyal agents of the central government, did. Letter after letter, full of multitudinous details, added to the crown's knowledge of the situation and helped form its decisions.

For example, Charles Machaut, from the army of Burgundy in 1636, reported his suspicions of Dijon's loyalty; the parlement and the other privileged groups were ready to foment sedition. Another army intendant, Argenson, described the condition of the army in 1636, noting that the *commissaires* had reviewed the regiments raised in Paris, but the troops lacked swords, muskets, and shoes. The new troops were ready to disband or mutiny unless the crown paid them. Meanwhile, enlarging a bridge to accommodate the passage of the cavalry took two hours before a corps of guards could advance to the other side. In another letter, Argenson spoke of the difficulty of maintaining proper sanitation in the army: a bureaucratic mix-up had occurred, and peasants hired to clean the camp could not be paid without signed authorization. In the future, Argenson advised, a fund should be set up to pay for this monthly expense. That same year another intendant, Aubray, reported that the army camp at Lussi needed to be put in order, the government's decision to lodge the troops in the countryside had been made too late, and the officers in charge brought more confusion than order to the task.[37]

36. Schmidt, "Le Rôle et les attributions d'un 'intendant des finances' aux armées," pp. 160–63; Machaut to Séguier, 14 September 1636, in Mousnier, *Lettres au Séguier*, 1:349–50; instructions to Argenson going to Catalonia, 18 February 1641, in José Sanabre, *La Acción de Francia en Cataluña en la pugna por la hegemonía de Europa (1640–1659)* (Barcelona, 1956), pp. 643–45.

37. Bassompierre's memoir, B.N. *cinq-cents Colbert* 499, fol. 99. Letter of Machaut, 14 September 1636; Argenson, undated (December 1636); Argenson, 1 November 1636; Aubray, 21 September 1636; all to Séguier, in Mousnier, *Lettres au Séguier*, 1:349–50, 338–39, 322–23, 309–10, respectively.

The intendant's importance to his superiors depended upon the accuracy of his information, for the government needed agents it could trust. It is therefore not surprising that the crown sometimes sent others to check on the intendants. The instructions of Le Tellier, upon his appointment as intendant of the army of Italy in 1640, ordered him to investigate the actions of his predecessor, Argenson, and of the *commissaire des guerres*, Talon, who had been denounced to Bullion, surintendant of finance, by two other *commissaires* and a controller. Le Tellier found no evidence to support the charges, and the government continued to use the services of Argenson and Talon, but the crown persisted in its demand for uncompromising loyalty and obedience from its agents.[38]

ARNAULD D'ANDILLY: INTENDANT IN ACTION

After examining an intendant's functions according to his commission and a few scattered letters, it is possible to reconstruct one such intendancy in action in the 1630s. Robert Arnauld d'Andilly, who left a precious memoir of his intendancy, was a member of the famous Arnauld family, destined from the cradle for government service. His father, a famed parlementarian and a loyal servant of Henry IV, saw that his son received an excellent education from private tutors, although it does not appear that Robert studied at a law school—a prerequisite for many government offices. Furthermore, there is no indication that Robert was a master of requests. Probably he did not need this office for entrance into the administration, because his uncle Isaac was an intendant of finance who took him on as one of his assistants. The young man seemed to have a promising career before him, for he won the favor of both the king and the king's brother, the duke of Orléans. In 1622 Louis XIII offered Robert the office of secretary of state on condition that he pay the family of his predecessor a hundred thousand *livres* in recompense. The proud Arnauld refused, perhaps because he thought that he would eventually get the office without paying for it. This never occurred, and Arnauld, compromised by his friendship with the scheming duke of Orléans, never rose above the second rank of government service. Toward the end of his career, Richelieu decided to appoint him army intendant in Germany.[39]

38. Caron, *Michel Le Tellier*, pp. xlvii–xlviii.
39. The memoir is printed in several collections. I have used Petitot, "Collection des mémoires relatifs à l'histoire de France," 2d ser., vols. 33 and 34, particularly 34:39–72. All subsequent parts of this chapter are taken from this memoir, unless otherwise noted. Biographical details on Arnauld are in Prevost and Roman d'Amat, *Dictionnaire de biographie française*, s.v. "Arnauld d'Andilly (Robert)."

In 1634 Arnauld received a letter from Abel Servien, the secretary of state for war, officially inviting him to assume the intendancy of the army commanded jointly by the marshals de La Force and Brézé. Arnauld tried to decline the honor in a private interview with Servien; he was too old to hope for higher posts and regarded the intendancy as hard work without much reward. Servien told him that Richelieu would be offended if he refused. Evidently Arnauld was still hesitant, for he next met with Père Joseph, the cardinal's confidant, who told him that he had no choice in the matter. Arnauld accepted and had an audience with Richelieu, who gave him his final briefing. "He told me, among other things, to live well with Marshal de Brézé, and he had recommended to him [Brézé] to use me in the same way." Finally, Arnauld had a ceremonial audience with the king before taking his leave of the court.

Arnauld also received a written set of instructions signed by both Bullion and Bouthillier, the joint surintendants of finance, in addition to the cardinal's verbal directions. Arnauld made no mention of Servien's instructions, and in his intendancy he was largely responsible to the two surintendants; the secretary of war did not yet play the predominant role that Sublet de Noyers and later Le Tellier did. Unfortunately Arnauld's instructions have not been preserved, and he included only a brief mention of them in his memoirs. He was, however, very proud of the fact that he was authorized to dispose of ten thousand *livres* per month for extraordinary expenses without having to account for it to the generals, "which I do not know to have been accorded to any other intendant of the king's armies." It is obvious that the government trusted Arnauld's maturity and judgment, for the power to spend such a large sum emphasized the intendant's independence from the generals.

Arnauld left Paris on 2 November 1634, and like most intendants, his first care was to provision the army. Arriving at Châlons-sur-Marne, which "is one of the cities of France where the most grain is found," the intendant contracted for a quantity of grain and transportation. Arnauld prided himself on his negotiations. He claimed to have saved the king two hundred thousand *écus* above the prices normally paid to the munitioneers, and he certainly appeared astute, for he took one measure of grain, ground it fine, baked it, and then weighed it to calculate how many rations of bread each measure of grain would furnish. He then used this as a standard for his purchases.

Arnauld proceeded to Nancy, where he joined Marshal Brézé and accompanied him to Lunéville, where the two came to an understanding. One of the gravest problems that any intendant faced was his relationship with the commanding officer. If the two did not get along or if they worked at cross-purposes, the entire expedition could end in failure, not to mention the harm that might be done to their careers. Generals such as Brézé were already apprehensive of an intendant's authority rivaling their own, and they were doubly suspicious of an agent such as Arnauld, who had a reputation for being haughty. The government always urged army intendants to live peacefully with the commanding officers, but not all intendants had the savoir faire that this task required.[40]

Brézé had a dramatic confrontation with Arnauld, whom the marshal summoned to his room one night. The marshal had heard rumors about Arnauld, and his friends had advised him to refuse the command of the army rather than accept Arnauld as an intendant, "because [in Brézé's words] you would like to act with so much authority that I could only be discontent." Whereupon the marshal enumerated certain advantages that he had over the intendant, such as being a marshal of France and Richelieu's brother-in-law, but Brézé conceded that Arnauld had advantages over him also and suggested that they forget such things and live together in harmony. The two concluded their bargain, and became the best of friends. Brézé wrote to the government that he had been given a veritable treasure in the person of the intendant, and Arnauld always spoke warmly of the marshal.[41]

After cementing his relations with Brézé, Arnauld discovered that feeding the troops remained his most important task. It was near the end of the year, and there were not enough funds to provide bread for the infantry. Because of his connections with the cavalry officers, however, he was able to arrange a temporary loan of forty thousand *livres* to tide the army over. But the problems of supply persisted. The government had entered the war without proper preparations, the winter that year was extraordinarily bitter, and food was difficult to obtain, but Arnauld somehow provided. He even envisioned a three-

40. Cf. also the instructions of Bullion to the newly appointed intendant of the army of Italy, Le Tellier, in 1640: "M. Le Tellier s'insinuera, le plus doucement qu'il lui sera possible, aux bonnes grâces de M. le comte d'Harcourt [the commanding officer]" (Caron, *Michel Le Tellier*, p. 4).

41. Georges Livet, *L'Intendance d'Alsace sous Louis XIV, 1648–1715*, Publication de la Faculté des lettres de l'Université de Strasbourg, fascicule 128 (Paris, 1956), p. 42*n*3, quotes from various archival documents to support this friendship between Brézé and Arnauld. Arnauld, however, did not get along very well with the other commanding officer, the marshal of La Force. In his memoirs, Arnauld tells the story that Richelieu was so content with the intendant's efforts that he wrote him a letter of commendation. The cardinal also wrote to the marshal de La Force, and somehow the addresses got mixed. La Force received the letter meant for Arnauld. "Ainsi il [La Force] vit ce que M. le cardinal me mandoit, et trouva que cette lettre étoit plus obligeante pour moi que celle qui étoit pour lui ne l'étoit à son égard" (Arnauld, *Mémoires*, 34:47).

zoned supply system whereby grain would be gathered in Champagne, stored in magazines in Lorraine, and then transported to Alsace, the theater of war.[42]

Illness also stalked the army during the severe winter, and six thousand soldiers required medical treatment. Arnauld cared for this large number, proudly remarking in his memoirs that "almost none of them died." At the same time, he castigated the primitive medical facilities in the army. The horses received better attention than the wounded, he claimed, for it cost more money to replace the animals than the men.

When the army besieged Spires, Arnauld remained behind at Landau "to give order to the things that are necessary for the subsistence of the army," and forwarded all the provisions he could find, together with medication for the sick and a quantity of wine to be distributed free to each soldier. At the conclusion of the siege, Arnauld brought the wounded back to Landau on litters and quartered them in nearby monasteries. The lot of the disabled veterans was a difficult one, so Arnauld announced that he was giving, in the king's name, a bonus of three *pistoles* apiece to more than two hundred of the most severe casualties.[43]

In his zeal for assisting the army, the intendant twisted regulations and got into trouble with the court. The king had decided to send some of Brézé's troops to Flanders to aid the Dutch, and the government desired that the troops be in exceptionally good condition for this expedition. The captains of the companies came to Arnauld protesting that this was impossible after such a severe winter. Only if the incomplete companies were paid on the basis of full units could the captains find the money to put their companies in first-rate condition, well armed and equipped. Arnauld agreed that this would be advantageous for the king's service, but he "had his hands tied, for he was not free to pay any but the number borne on the records of the reviews." The captains were adamant, Brézé added his appeals to theirs, and Arnauld wavered, although he knew that nothing was more extraordinary than to pay for more men than were listed in the reviews. Finally Arnauld capitulated and paid the captains for the imaginary soldiers. The captains used their money wisely, and their men were the "plus belles troupes" that anyone had ever seen.

42. Although this project was never adopted, Arnauld's intendancy, according to Livet, "représente un étape dans la prise de conscience par la monarchie administrative des nécessités d'une guerre européenne" (Livet, *Intendance d'Alsace*, p. 43).

43. According to Marion, *Dictionnaire des institutions de la France*, s.v. "Monnaies," the *pistole* was equal to ten *livres*.

Shortly thereafter, Arnauld received orders to appear at court. "I was not a little surprised that in place of signs of satisfaction that I had expected of him, he, Bullion [the surintendant of finance], began by quarreling with me, telling me that I had paid the troops more than the amount authorized by the reviews." Arnauld tried to defend his action on the grounds that it was for the good of the king's service, but Bullion refused to listen. Arnauld then sought an audience with the cardinal and, Servien being present, accused Bullion of false accusations. Arnauld dramatically pulled the financial accounts of the army from his pocket to prove his probity. The cardinal was astonished. Of the ten thousand *livres* per month at his disposal, he had not spent half—despite the expense of the hospital and the bonuses to the wounded at Spires. The cardinal accepted his excuses, and Arnauld returned to the army.[44]

In justification of the intendant's conduct, it should be mentioned that Arnauld was renowned for his integrity, and his reputation helped excuse his transgression. Nevertheless, it was extraordinary when an intendant took it upon himself to judge that necessity merited disobeying orders without consulting his superiors at court. Arnauld behaved as an independent agent in the field with little regard for the wishes of the war department. This would change when Le Tellier became war secretary; no agent would dare take such liberties with him.

Arnauld returned to the army, but his career was anticlimactic after this event. Sick much of the time, unhappy with changes in the army, and tired of his burden, Arnauld asked for his own recall. He received permission to return to court at the end of the 1635 campaign. There he paid his respects to the king before rendering an account to Richelieu of all that had occurred during his intendancy, including the conduct of the officers.

Certain conclusions can be drawn from Arnauld's intendancy. First, the war department played little role in Arnauld's story. Bullion, the surintendant of finance, had more power over Arnauld than did Servien, the secretary of state for war. As his title indicates, Servien's job was that of a secretary to the king who handled correspondence on military affairs. It was only under his successors, Sublet de Noyers and Le Tellier, that the war department with its permanent clerks and secretaries took form. Although powerful ministers like Richelieu and Mazarin continued to control the appointment of army intendants, there would be a time when Le Tellier and his son Louvois would con-

44. See also in this context the memoir of Brézé, "Articles de memoire de M. le surintendant auquels iay respondu," A.A.E. *FRANCE* 813, fols. 63–64.

struct their own team of intendants, bound by ties of blood and loyalty and obedient to the will of their master, the secretary of state.[45]

Second, concern for provisioning the army slowly replaced concern for matters of justice. In the future, an intendant's legal training was less important than his administrative experience. Arnauld was a prototype. He had no legal expertise, and he did not administer justice in the army; his goal was to supply the army. "I was persuaded that the greatest service that I could render," Arnauld remarked at the beginning of his intendancy, "was to work with all my power for the army's subsistence." In this matter, the crown had granted Arnauld full authority: "My charge gave me an entire power over the officers of supply." Provisioning would remain the chief care of future intendants.

Finally, an army intendant might be a powerful support—or a threat —to the general, for he had the ear of the government. When Arnauld returned to court, he rendered an account of his mission, recording an anecdote that illustrates how powerful an intendant's evaluation might be. After his final interview with Richelieu, Mazarin banteringly said to the intendant: "M. d'Andilly has spoken of M. de Feuquières [a lieutenant general] to the cardinal in such a manner that he has today made his fortune, because . . . this discourse has made such an impression on his [the cardinal's] mind, that it is impossible that M. de Feuquières does not feel its effects." Although this is perhaps another of Arnauld's boastful stories about how much others owed to his friendship, it does hold a moral: few army officers were powerful enough to offend an army intendant. Usually the intendant and the commanding officer lived together with mutual understanding; but when they did disagree, the government had problems. This situation would continue to plague the government throughout the days of the ancien régime.

45. See Ranum, *Richelieu and the Councillors of Louis XIII*, chaps. 3 and 5, for a brief survey of the evolution of the war department and the secretaryship of war. Ranum suggests that historians have too long ignored the role of Sublet de Noyers and overstressed that of Le Tellier.

IV

The Intendants under Le Tellier
1643-48

With the death of Richelieu on 4 December 1642 and the passing of the king barely six months later, the French government faced a change in personnel. On the day after Richelieu's death, Louis XIII called the cardinal's chosen successor, Mazarin, into his councils— a man in whom, the king declared, he had full confidence. Mazarin retained his position after the king's death and gained the trust of the regent, Anne of Austria, mother of the young Louis XIV. Cardinal Mazarin quickly replaced the secretary of state for war, Sublet de Noyers, with his own appointee, Michel Le Tellier, the intendant of the army in Italy. Mazarin announced the appointment to the young man in these words: "You will not doubt that I have only rendered the testimony that the truth and my affection for you oblige me to render," and concluded: "as for the passion that I have for what touches you, time and its effects will teach you better than my words." The cardinal had made an excellent choice, for Le Tellier became Mazarin's faithful collaborator and served him obediently all his life.[1]

1. Pierre-Adolphe Chéruel, *Histoire de France pendant la minorité de Louis XIV* (Paris, 1879), 1:3-4, 23-24. Mazarin to Le Tellier, from Saint-Germain-en-laye, 11 April 1643, in Pierre-Adolphe Chéruel and Georges d'Avenel, eds., *Lettres du cardinal Mazarin pendant son ministère*, Collection de documents inédits sur l'histoire de France (Paris, 1872), 1:152-53. Many rumors circulated concerning Sublet's disgrace. Some believed that it was due to his friendship with the Jesuits, others that he had had absolute disposition of more than thirty million *livres* under Richelieu and now had refused to account for it to the king. Mazarin, however, had made his plans well in advance and had convinced the king "qu'il [Le Tellier] estoit capable d'un plus grand employ." As was usual at the court, an official's fall from favor meant that everyone turned against him even though one might owe him gratitude for past favors. Le Fèvre d'Ormesson, for example, recounted that the day after Sublet's disgrace, the chancellor, "who owed his fortune and his protection to M. de Noyers," refused to seal a commission for M. de Champigny-

There resulted a general shuffling of personnel. A few, such as Isaac Laffemas, were so odious to the populace that the government gained popularity by exiling them to their estates. But not all officials were exiled; some were kicked upstairs: "The grand door of the Council opened and all who desired it, entered therein," wrote Olivier Le Fèvre d'Ormesson in his *Journal*, "and there was great confusion." After 1643, the names of many who had served as army intendants, such as François Lasnier, Gaspard Du Gué, Claude Gobelin, Charles Machaut, Louis Le Maistre de Bellejamme, and Alexandre de Sève, gradually disappeared from the list of military intendancies and were replaced by a flood of new ones: Champlastreux, Balthazar, Fouquet, Brachet, Goury, Clermont, and Baussan.[2]

It is difficult to determine precisely whose clients these were, yet tantalizing hints remain for the careful observer. Mazarin took a special interest in the career of Jean-Edouard Molé, sieur de Champlastreux, perhaps because of the young man's family connections—his father was a president of the Parlement of Paris. Another of the cardinal's confidants was Jean Balthazar, who served as intendant of Languedoc from 1643 to 1647, and then in the army in Lombardy. Mazarin wrote frequently to Balthazar, mixing affairs of state with personal news such as congratulating him upon the birth of twins and mentioning the queen's willingness to become his daughter's godmother. And one

Nauroy as intendant of Pignerol—a request the king had granted Sublet before his downfall. The chancellor erased Champigny's name and substituted another. Pierre-Adolphe Chéruel, ed., *Journal d'Olivier Lefèvre d'Ormesson*, Collection de documents inédits sur l'histoire de France (Paris, 1860), 1:23–25, 33.

2. Laffemas was exiled to Issoudun in March 1643. On his way into exile, he was robbed on the route by his enemies; Chéruel, *Journal d'Ormesson*, 1:15, 18. The quotation from the journal of Le Fèvre d'Ormesson in the text is in Marion's *Dictionnaire des institutions de la France*, p. 131; Marion attributes this remark to vol. 1, p. 424, of the *Journal* edited by Chéruel, but I could find only part of Marion's extensive quotation on this page, not, however, the passage quoted above.

There is some disagreement about whether Mazarin's regime meant a house-cleaning of government officials. Jules-Auguste Lair, *Nicolas Fouquet, procureur général, surintendant des finances, ministre d'état de Louis XIV* (Paris, 1890), 1:83, remarked: "contrairement aux espérances des mécontents, la Reine et son ministre laissèrent en place et même favorisèrent les 'créatures' de Richelieu." Neverthe-less, there was a turnover in personnel. Not only did the government drop Sublet de Noyers as secretary of war, but the important Bouthillier family also suffered reverses. Claude le Bouthillier, superintendant of finance whose star had risen through his relationship to Richelieu, bore alternate periods of favor and disfavor before being forced to retire to his estates after Louis XIII's death. His son, Léon le Bouthillier, count de Chavigny and secretary of state for foreign affairs, was so dependent upon Richelieu that his own power began to wane on the cardinal's death. He died, disgraced, in 1652. See Ranum, *Richelieu and the Councillors of Louis XIII: A Study of the Secretaries of State and Superintendents of Finance in the Ministry of Richelieu, 1635–1642* (Oxford, 1963), pp. 99, 180.

must not forget Nicolas Fouquet, surintendant of finance, who rose to prominence under Mazarin's tutelage—serving as intendant of the army of Picardy in 1647 under the command of the duke of Orléans, and later in the armies during the Fronde.[3]

Le Tellier also sought to place his own confidants in positions of trust. In 1645 he announced to Girolles, the frontier intendant of finance, fortification, and supplies in Brisach: "My Lord the cardinal has found it good that M. de Baussan, my close relative, should have the intendance of Alsace. I am his surety and assure myself that his conduct will content everyone." Girolles was also a relative of Le Tellier, and had served as the secretary's assistant prior to his appointment in Brisach. Another of Le Tellier's cousins, Charles Brèthe de Clermont, joined the administration at this time, having the direction and intendance of the troops in the Boulonnais in 1644; later he became intendant of the frontier places of Picardy and Artois (1646), and of Arras. Pierre Goury, another relative, worked closely in the war department with the secretary before serving as intendant of the army in Catalonia from 1645 to 1647. As Le Tellier secured his position, his influence in government widened, and many more friends and relatives joined the administration, clients of a client of Mazarin.[4]

Before examining the intendancies of these men, one must remember

3. There are a number of Mazarin's letters to Champlastreux summarized in the "Table chronologique des lettres analysées" in each volume of Chéruel's *Lettres du cardinal Mazarin*. For example, in the second volume there are felicitations on the manner in which Champlastreux conducted himself in the army (2:614), protestations of affection (2:684), praise of Champlastreux's conduct and his handling of the king's money (2:625), and Mazarin's response to Champlastreux's thanks for naming his brother bishop of Bayeux (2:924). There are similar letters of Mazarin addressed to Balthazar: congratulations upon the birth of twins while acknowledging the treaty made in Catalonia for furnishing bread to the army (2:714), thanking him for a visit to brother Michel Mazarin and for services rendered to Cardinal Barberini (2:828), and announcing his army intendancy and praising his person to the commanding general, Du Plessis Praslain (3:999).

Nicolas Fouquet began his career under Richelieu's protection, being named "intendant de police, justice et finances auprès de l'armée chargée de deffendre la frontière septentrionale" in 1642. After Richelieu's death, he continued, under Mazarin's protection, as intendant of Dauphiné and army intendant before rising to the position of surintendant of finances. Lair, *Nicolas Fouquet*, 1:78–84. Pierre-Adolphe Chéruel, *Mémoires sur la vie publique et privée de Fouquet surintendant des finances d'après ses lettres et des pièces inédites conservées à la bibliothèque impériale* (Paris, 1862), 1:5–8.

4. André, *Michel Le Tellier et l'armée*, p. 636n4. Le Tellier's letter to Girolles is also printed in Georges Livet, *L'Intendance d'Alsace sous Louis XIV, 1648–1715*, Publication de la Faculté des lettres de l'Université de Strasbourg, fascicule 128 (Paris, 1956), pp. 96–97. André claims that Le Tellier had close relations with Champlastreux—"avec qui il est en relations suivies" (*Michel Le Tellier et l'armée*, p. 636)—but Mazarin's relationship with Champlastreux seems more intimate (witness his letters to Champlastreux in the preceding note). I found no such warmth in Le Tellier's letters.

a basic fact. In the five years between Le Tellier's entrance into office (1643) and the outbreak of the Fronde (1648), his position was extremely tenuous; consequently no great or wide-sweeping military reforms took place. Two years after his appointment, Le Tellier still had only the title of provisionary secretary of state. Sublet de Noyers had resigned the office, but he retained a legal title to it and delayed Le Tellier's attempts to buy the office, by raising the price in hope that if he delayed long enough, he might return to favor. This was not to be, but Le Tellier did not gain permanent possession of the office until after Sublet's death on 20 October 1645.

Three years later the Fronde broke out and Le Tellier's position once again became insecure. Because he was hated as a client of Mazarin, Condé and several princes of the blood demanded his ouster along with other officials. The secretary had no choice but to retire to his estates at Chaville to wait out his misfortune, while Loménie de Brienne handled the affairs of the war department for six months. Late in 1651 a more ominous threat surfaced when Mazarin expressed suspicions of Le Tellier's loyalty and feared that the secretary sought to replace him as first minister. Fortunately for Le Tellier, he was reconciled to the cardinal and remained secretary of state. Only after the Fronde was crushed and the young king crowned at Rheims was Le Tellier secure enough to begin far-reaching reforms in the army.[5]

While the office of army intendant responded to the reins of a new master, the institution did not drastically alter its form. It had always been responsive to the wishes of the government, and now it matured and broadened its authority, constituting the government's trusted weapon in the army.

One of Le Tellier's main concerns as secretary of state was to learn the actual condition of the army, for his days as intendant of the army in Italy had taught him that a civilian chief needed daily information if he hoped to administer the army effectively. Almost from his first day in office, he kept officials busy furnishing statistics, accounts of army expenses, rolls of troops, receipts of contributions, and reports on the condition of magazines. "Render me an account of what you have done," "you will send me a copy as soon as you have drawn it up," "the rolls . . . are to be drawn up very carefully and sent to me as soon as possible," "send me the duplicate"—these phrases appear over and over in Le Tellier's correspondence.[6]

---

5. André, *Michel Le Tellier et l'armée*, pp. 89-113, especially pp. 111-12, regarding Le Tellier's ability to begin reforms only after 1652.

6. The king to Ennemond Servien, intendant of the army of Italy, 17 July 1648, asking him to remonstrate with the Savoyard court concerning the levy of customs taxes on supplies shipped by munitioneers: "et que vous me rendies compte

Army intendants recognized the secretary's request for information and were prompt to reply. "I know that you always will be at ease," wrote Pierre Goury after sending an account of a troop review, "to be promptly informed, so that in the plans that the court may order, you can clearly judge the number of troops available for the campaign." Other memorandums, rolls, and accounts flowed in to Le Tellier. "I am sending you herewith the receipt which I have received," wrote Servien from Italy, while Goury responded in Catalonia: "I am enclosing a copy of the statement of the sum expended for the siege works, signed by me." Examining Goury's correspondence for the month of August 1648 shows that he sent Le Tellier the following memorandums: a statement of the quantity of grain captured in the city of Tortosa, together with a promise to send an exact account of all the grain, flour, biscuits, and other food supplies recovered; a receipt for the suits, shoes, and hats that Le Tellier had forwarded to the army; an estimate of the expenses involved in the siege of Tortosa and a notice of his intention to send, as soon as possible, more precise figures; and a memorandum of the number of troops, both cavalry and infantry, in the army and garrisons of Catalonia as revealed by the last review.[7]

Because the intendant played such an important role in describing

de ce que vous aves faites en consequence de la present," A.G. A[1] 108, pp. 17r–18r. The king to Charles Brèthe de Clermont, intendant of the army on the Flanders frontier, 22 July 1647: "que vous fassies les etats par[ers] [*particuliers*] en chaque bailliage ou chastellenie de cequi devra y estre levé en argent et denrées et sur iceux un estat general duquel vous vous menvoyeres le double aussitot que vous l'aures dressé," A.G. A[1] 104, pp. 38v–40v. The king to Servien, 10 December 1647, concerning winter quarters in Italy: "quil fasse dresser un rolle bien distingué par regiment et compagnies des noms et qualités des officiers qui demeureront en Italie, et un autre de ceux qui reviendront dans le royaume, desquels il faudra que le double me soit envoyé," and later in the same letter: "prenies soin que les rolles que je mande a mond. cousin, des officiers qui demeureront en piemont de ceux qui reviendront aux recreues et de ceux de cavallerie qui ont este absents de leurs charges soient tres exactement dressés et me soient envoyés aussitot quil se pourra, comme aussy que je sois bien particulierement informé de la force veritable des troupes qui demeureront," A.G. A[1] 104, pp. 284v–88r.

Although these letters are formally from the king, a young boy born 5 September 1638, they are really those of Le Tellier acting under the authority of the regency government and are bound as the "despeches" (*sic*) of Le Tellier in the war archives at Vincennes. The regency government lasted from Louis XIII's death in 1643 until Louis XIV's majority in 1651. The queen mother, Anne of Austria, served as regent, and documents of the period usually contained a phrase, such as the following, acknowledging her authority: "par l'advis de la reyne regente madame ma mère." In chapters IV and V of this study, I have simply referred to Anne as the queen.

7. Goury to Le Tellier, 12 August 1648, A.G. A[1] 106, fol. 132; Ennemond Servien to Le Tellier, 6 June 1648, A.G. A[1] 110, fol. 14; and Goury to Le Tellier, 3 and 12 August 1648, A.G. A[1] 106, fols. 123, 132.

the condition of the army, the secretary of state never allowed an intendancy to remain vacant. Even when an intendant was absent for a short period, Le Tellier always sent a substitute. In 1647, for example, the secretary named Pierre Imbert, intendant of Roussillon, as temporary replacement for Champlastreux, intendant in the army of Catalonia, although a permanent army intendant would be named three months later. That same year, Le Tellier appointed René Le Vayer to exercise the intendancy of the army on the northern frontier commanded by the marshals Gassion and Rantzau until the regular army intendant, Nicolas Fouquet, arrived to take up his intendancy.[8]

Although Le Tellier's experience as intendant had taught him the importance of gathering statistics, another circumstance made it essential that he know everything that happened in the army—its expenses, the number of troops, their condition. France had been at war since the 1630s, and her financial organization proved incapable of meeting the high cost of continual warfare. The government resorted to heavier taxes and used its own agents, the provincial intendants, to insure their collection. The result was tax rebellion, both by the peasants in popular uprisings and by the treasurers of France and other fiscal agents who felt themselves threatened by the crown's usurpation of their functions and resorted to sabotaging tax collections. The government found itself less and less able to send money to its armies. Le Tellier carefully examined all requests for funds, cut expenditures, and resorted to various expedients to scrape up the income. Consequently it was important for the secretary to know exactly how much money the armies spent and what the government got in return.[9]

8. The king to Imbert, to assume the army intendancy vacated by Champlastreux, "lequel il importe de ne laisser aucunement vaccant," 12 September 1647, A.G. A¹ 104, pp. 96v–97v. The employment was clearly temporary, for the letter continued: "voulant que le pendant vous fassies toutes les fonctions de celle d'intendant des finances comme si vous en avies le titre et ce jusques a ce que je vous en aye deschargé apres que jauray pourveu a lad. intendance." Three months later, on 24 December 1647, Pierre Goury obtained a commission as "intendant des finances et fortifications en notred. armee et places de Catalongne," to replace Imbert, A.G. A¹ 104, pp. 314v–16v, and Imbert received a letter of recall from Paris dated 23 December 1647, A.G. A¹ 104, pp. 317v–18r. Regarding Le Vayer, see letter from the king, 23 May 1647: "et le Sʳ Fouquet que jay choisy pour faire la charge dIntendant en mad. armee n'ayant encore pü sy rendre. . . . vous y rendre pour faire les fonctions d'intendant. . . . et lorsque led. Fouquet sera arrivé en mad. armée vous pourres retourner a Arras," A.G. A¹ 103, p. 354rv.

9. Louis André does not give precise figures for the cost of the war from 1643 to 1659, remarking: "De 1643 à 1659, c'est-à-dire pendant la guerre, l'évaluation ne peut être qu'approximative: car le nombre et l'effectif des régiments d'infanterie et de cavalerie changent à tout instant et dépendent en même temps des complications qui surgissent à l'intérieur." But he estimates the cost from 1653 to 1655 at approximately 11 million *livres* for a two-month period, or 66 million for a

Matters grew worse as the country entered the period known as the Fronde. Le Tellier received daily pleas for funds from all military fronts. In Italy, the general Du Plessis Praslain wrote: "the extreme necessity by which we are embarrassed is such that we have not one day's worth of flour left in the army"; Goury, in Catalonia, reported that he was continually pressed for money, and warned that unless the officers were paid they would return home. The army on the northern frontier of France faced similar problems; the munitioneer lacked funds, and even the intendant had five thousand *livres* due him.[10]

Sacrifices had to be made on all sides. Le Tellier wrote Du Plessis Praslain that the government was doing all it could to find money; "also My Lord Cardinal, through the credit of his friends, makes extraordinary efforts for your assistance." Le Tellier ought to have counted himself among these friends, for he had lent the government some three hundred thousand *livres* from his own resources. The generals and the army intendants made similar sacrifices. For example, Du Plessis Praslain announced from Italy: "We have put all that we have in hock in Mantua. M. de Modena [the duke] contributed what was possible on his part, M. Balthazar also, and M. Brachet [the intendants]. I have given my dinner service and all that I have of silver, as much for siege work as to get grain." [11]

During this period of crisis, the intendants did everything they could to help the army scrape through. Le Tellier warned Servien, intendant

year. André, *Michel Le Tellier et l'armée*, pp. 291–92. Elsewhere (p. 271) André stressed that the financial question was closely tied to the military one, "la détresse de l'une entraîne fatalement la détresse de l'autre."

10. Du Plessis Praslain to Le Tellier, 16–17 August 1648, from the camp before Cremona, A.G. A¹ 106, fol. 140. Goury to Le Tellier, 20 August 1648, from Barcelona, noting the need to pay the troops and the munitioneer; otherwise "il seroit a craindre que les armes du Roy ne receussent un grand eschecq en ce pays par un desbandement general qui surviendrait infailliblement parmy les troupes," A.G. A¹ 106, fol. 143. Clermont to Le Tellier, 5 August 1648, from Calais, A.G. A¹ 106, fol. 125. Almost every one of Goury's letters in 1648 was an appeal for additional funds, although France had poured more money into the army in Catalonia than any other. See the crown's instructions to the previous army intendant, Champlastreux, in 1647, which noted: "quil ny a aucune armée pour laquelle Sa Ma^te fasse tant de despence que pour celle de Catalogne," A.G. A¹ 103, pp. 172r–84r.

11. Le Tellier to Du Plessis Praslain, 13 June 1648, A.G. A¹ 107, pp. 288r–89r. André, *Michel Le Tellier et l'armée*, p. 277. Letter of Du Plessis Praslain, 16–17 August 1648, A.G. A¹ 106, fol. 140. Du Plessis Praslain's memoirs agree: "Tout ce qu'il avoit d'argent y fut employé, tout celui de ses amis de l'armée y fut consommé de même; et enfin, n'ayant plus d'autre ressource, il vendit sa vaisselle d'argent: tellement qu'il employa du sien en cette campagne environ quatre cent cinquante mille livres pour la nourriture de l'armée," César Choiseul du Plessis Praslain, *Mémoires des divers emplois et des principales actions du maréchal du Plessis*, ed. Michaud and Poujoulat, Nouvelle collection des mémoires relatifs à l'histoire de France, vol. 31 (Paris, 1857), p. 397.

of the army in Italy: "You will have to contribute everything in your vigilance and attention to conserve all the supplies." Servien responded by reducing the ration of bread in the army, although he delayed this measure until the first of July so that the troops would not desert while on march, promising, "I will manage the best that I can to see that six thousand loaves suffice for all the troops we have in this campaign." At the same time, munitioneer Falcombel and his brother warned the intendant that they would be forced to discontinue furnishing bread within two weeks unless Servien paid the money owed them. Le Tellier tried to save the situation in Italy by retiring some of the troops to France, but despite this relief, money was still lacking. At a council of war attended by Cardinal Antoine, the marquis de Ville, and Saint André, Sevigné, and Servien, the officers proposed that each of them advance the sum of 30 *pistoles*, making a total of 150 *pistoles*, so that the army temporarily might survive.[12]

Pierre Goury resorted to similar tactics in Spain. He reduced the amount of bread given to the troops, borrowed money, and refused to accept the sick into the hospitals. Goury cut the amount of bread furnished to the army to 14,692 rations, remarking: "I have brought all my attention to bear for the management of the finances and for the reasonable satisfaction of the troops." In August 1648, Goury had not yet received the funds for the army allocated for the month of July. He went into debt for more than sixteen thousand *pistoles* for grain and various extraordinary expenses, but his credit soon ran dry, and by the end of August he remarked that he had difficulty finding anyone to lend him a thousand *pistoles*. Having no more funds for the hospital, he left the medical attendants at the captured fortress of Tortosa with orders not to receive any more patients unless the army engaged in active fighting, adding that "it is completely impossible to furnish them anything," since he could not find the funds.[13]

Given these conditions, there was little else the government could expect the army intendants to do, but the government's hopes were expressed in detailed instructions at the beginning of each intendancy. Champlastreux's instructions in 1648—one of the few still preserved in the archives—deserve critical attention. They begin with a warning for the intendant to act under the authority of the prince of Condé, the commander in chief of the army and viceroy of Catalonia, but

12. Le Tellier to Servien, 10 December 1647, A.G. A¹ 104, pp. 284v–88r; Servien to Le Tellier, 6 and 13 June 1648, A.G. A¹ 110, fols. 14, 42. Cardinal Antoine to Le Tellier, 10 June 1648, A.G. A¹ 110, fol. 32.

13. Goury to Le Tellier, 3, 12, and 20 August 1648, A.G. A¹ 106, fols. 123, 132, 143.

they proceed with this injunction: "In regard to finances, he [Champlastreux] will examine all the statements and ordinances which will be decided upon and authorized by My Lord the Prince, resulting from the orders of His Majesty, for the expenses of the war in Catalonia. He will verify the statements of these payments, month by month, and send a copy to His Majesty to inform him punctually of the use of the funds which will have been sent to Catalonia and of the sums remaining." [14]

This clause underlined a constant theme: the intendant was the fiscal watchdog of expenses in the army. Le Tellier was serious about this matter, and he wished it understood "that there is nothing more important for the service of His Majesty than to work for the good management of his funds . . . His Majesty wants him to be well informed about all that concerns the finances and whatever can contribute to conserve and employ them usefully in these quarters." [15]

Champlastreux's instructions suggested how this injunction was to be acted upon. The funds for the army were to be furnished in *pistoles* or *louis d'or*, both gold coins. This precious metal in war-torn Catalonia experienced an inflation of some 63 percent above its real value in France, for a *louis d'or* worth 10 *livres* in France was equal to 16 *livres* 6 *sols* and 3 *deniers* in Catalonia. The crown desired that its soldiers be paid in this "hard" currency, but the troops did not benefit from the full 63 percent inflation, for the value was partially discounted by the government. Nevertheless, it meant a substantial advantage for the common soldier. The intendant's task was to see that the troops benefited from this inflation by insuring that the men were paid in coinage sent from France. This meant that the payment was to be made in the intendant's presence, because he had to watch the agents from the treasury of war to see that they did not keep the hard currency for their own profit and pay the soldiers in worthless Catalan money. As additional insurance against fraud, each wagonload of coinage had a special invoice of its contents ready for the intendant's inspection.

Although this coinage was intended for payment of the troops, certain "necessary and pressing" expenses had to be paid before salaries, such as boats for pontoon bridges to cross the Spanish rivers, oats for the cavalry, transporting grain and munitions, siege works, and hospital expenses. Since the other expenses were to be paid in the debased

14. Instructions for Champlastreux as army intendant in Catalonia, 20 March 1647, A.G. A¹ 103, pp. 172r–84r.

15. This and the following paragraphs are taken from Champlastreux's instructions cited in the preceding note.

Catalan coinage, the intendant had to watch the officials of the treasury carefully to prevent them from reaching collusive agreements with local businessmen, paying them in hard currency and then splitting the profits—"as has happened in the past in Catalonia," the instructions added.

Champlastreux's instructions warned him to watch for subterfuge on the part of army officers as well as treasury agents. These officers claimed they owned several horses in the army and collected rations for their fictitious beasts, selling the grain on the side to make a nice profit. Le Tellier instructed the intendant to see that oats were given only to horses actually in the service, that is, those listed in the military reviews. Furthermore, army officers were not allowed to appropriate government property for their own use, particularly horses employed in the artillery. Officers in the infantry had a different scheme for cheating the government. They inflated the number of soldiers actually enrolled in their companies, Champlastreux's instructions warned, and received their bread rations, which they then sold for profit. Other officers received bread for themselves and their servants, although they knew that only soldiers and sergeants were supposed to get these rations. It was the intendant's duty to suppress such abuses.

The war department concluded its instructions by warning Champlastreux to watch carefully the army's expenditure, for the army should not expect an increase in funds during the campaign. If some unforeseen misfortune occurred and the funds did not suffice, Le Tellier left it to the prudence of the intendant and the general to deduct the needed money from the soldiers' pay. The intendant had to be careful in this, the instructions added, for the troops in Catalonia had to pay for everything furnished them from the countryside in order to preserve the goodwill of the natives; if the intendant reduced the pay, the soldiers would be in danger of perishing or reduced to pillage. In any case, if the intendant had to subtract money from the pay, it should not be done while the troops were in active combat.

It is obvious from these instructions that the intendant acted as an accounting agent, watching expenditure and trying to spot abuses. From watching expenses to controlling them was a short step, a step the intendants soon took. Although according to military practice only the general could order expenses, and order them usually on the basis of instructions from the king, the intendant had to countersign such documents. This was the secret of the intendant's power. Real control of expenditure soon passed from the general's to the intendant's hands during Le Tellier's administration.

This fact is apparent from an examination of the instructions of the prince of Condé, who accompanied Champlastreux to the army. His instructions discussed military strategy, the proposed sieges of Tarragona, Tortosa, and Lerida, and the hoped-for advantages from each. No mention was made of financial matters. The instructions told Condé: "His Majesty has specifically informed the sieur de Champlastreux, intendant of justice and finance in the army, of the funds that he [His Majesty] orders for the expenses therein, which forestalls one from saying anything in particular to the said lord, the Prince, assuring him only that they [the funds] will be delivered punctually." [16]

Since the general did not receive any additional financial information, Le Tellier continually reminded Champlastreux to keep Condé informed; "in order that the said Sieur intendant can give advice to My Lord the Prince conforming to the intentions of His Majesty," "sieur de Champlastreux must explain," "the said sieur de Champlastreux must employ himself near My Lord the Prince to see"—such were the phrases in Champlastreux's instructions. At the same time, Condé's instructions make no mention of the important fact that no additional funds would be sent to Catalonia. Surely this information was important for the commanding officer, if he actually was the one who ordered expenses. But Condé's instructions say nothing of this, while Champlastreux's stressed this information.[17]

Champlastreux was not exceptional. Other intendants also obtained control over an army's expenditure. In 1648 the king sent a letter to Condé's replacement, Marshal Charles de Schomberg, announcing the departure of 195,000 *livres* for the army in Catalonia. This money was to pay a quarter's wages to the garrison at Roses and to reimburse the communities and private individuals who had advanced money to feed the soldiers during the last campaign. The king asked Schomberg to order the payment of the troops "on the basis of the statement which I have sent, a duplicate of which I have addressed to Sieur Goury." Regarding the payment of the communities, Schomberg was to order this expenditure, "which will be countersigned by the said Sieur Goury," following the previous instructions of 16 November 1648. According to these instructions, the communities and individuals had to produce their receipts in proper form to intendant Goury, who was to pay them "following the verification that I order Sieur Goury to make." In such circumstances it was really the army intendant who was responsible for payment, although the general had the official pre-

16. Instructions given to the prince of Condé, Louis II of Bourbon "le Grand Condé," as commander of the army in Catalonia, 6 April 1647, in the appendix of José Sanabre, *La Acción de Francia en Cataluña en la pugna por la hegemonía de Europa (1640–1659)* (Barcelona, 1956), pp. 684–91, particularly p. 688.

17. Champlastreux's instructions, A.G. A¹ 103, pp. 172r–84r.

rogative of issuing the formal orders. The intendant had separate instructions from the government, and Goury did not need to rely upon the general for information.[18]

The balance between the general's and the intendant's power always remained a delicate one. In their instructions, the government told the intendants to act under the general's authority, and generals always wished to keep the initiative for themselves. Some, such as Du Plessis Praslain in Italy, pretended as if they determined all expenditure in the army. Du Plessis Praslain observed in a letter to the secretary of war in 1648: "The money you have sent will be managed as carefully as you might desire, and if M. Balthazar has written to you about the severity with which I have reduced the bread for the infantry, you can indeed judge that I was not afraid to embroil myself with its officers." While claiming credit for himself, the general conceded Balthazar's authority in almost the next sentence. The army had indebted itself to the extent of twenty thousand *livres* for bread that Sieur Cheneuix and Baron Cochet of Mantua had furnished. When they pressed Du Plessis Praslain for payment, the general admitted to Le Tellier, "M. Balthazar found it appropriate that I undertake to let you know that this sum is legitimately due them and to beg you at the same time to give it to them," adding at the end of his letter: "you will see from what M. Balthazar informs you, for I leave this subject to him." [19]

Ennemond Servien, another army intendant in Italy, provides an additional example of an army intendant's control of expenditure. In a letter to Le Tellier dated 13 June 1648, Servien remarked, "I loaned a hundred *écus* [to Brasset, provost of the army] to remount some of his archers, all of whom, save five, were on foot, having lost their horses in the last . . . campaign." In another letter, Servien told of offering one piaster to every Spanish soldier who deserted to the French forces, as Le Tellier had ordered. There are other examples from different intendancies during that same year. Although a *maréchal de camp* named Boissac had Le Tellier's approval for intendant Balthazar's payment of thirty-two thousand *livres* for the eight companies of his regiment, the intendant refused to pay this sum. Balthazar claimed that Boissac had furnished only fifteen thousand *livres* worth of recruits and did not deserve the additional sum. Consequently Boissac beseeched Le Tellier to countermand the intendant's actions: "I beg you . . . give the necessary orders for me to receive the remaining seventeen thousand *livres*." In one further example, Goury wrote

18. The king to Schomberg, 30 and 16 November 1648, respectively, A.G. A^1 108, pp. 184rv, 170r–72v.
19. Du Plessis Praslain to Le Tellier, 1 July 1648, A.G. A^1 106, fol. 95.

from Catalonia that the marquis of La Farre and of Saint Abre had asked the intendant to pay them the bonuses accorded them by the king for the months of May and June. Unfortunately, Goury reported, "I could not satisfy them because of lack of funds."[20]

The personal role of the army intendant was not limited to expenditure. The army intendants also continued to consolidate their control over areas of military administration: watching agreements with munitioneers, buying grain for the cavalry, supervising the transport of supplies and munitions, distributing clothing to the soldiers, and caring for the hospitals. Throughout the 1640s and 1650s the intendants gained expertise in handling these affairs.

The basic army ration, bread, continued to be contracted out to munitioneers, particularly Falcombel in Italy, Jean Martin in Catalonia, and a Monsieur Fly along the northern frontier in 1648. Too often the army intendants faced the problem of contractors who threatened to stop furnishing the army because they had not been paid. In such cases the intendant tried to use his limited powers of persuasion or credit to stall the munitioneer while appealing to Le Tellier to send funds quickly. The munitioneer Falcombel complained that unless he was paid within two weeks, he would quit furnishing bread to the army in Italy. Goury had similar problems in Catalonia. He had made an agreement with Jean Martin to supply grain to the cavalry. This contract continued during Champlastreux's tenure as intendant and still functioned after Goury's return to the intendancy in 1648. But Goury now had difficulty in paying Martin. Thanks to a bill of exchange from Le Tellier late in the summer of 1648, the intendant had succeeded in paying off some ten thousand *pistoles* due the contractor, but he still remained in debt for more than six thousand *pistoles* to Martin and others, which, he wrote to Le Tellier, was impossible for him to pay unless the secretary sent him additional funds.[21]

Despite Goury's problems with the munitioneer, he succeeded in

20. Servien to Le Tellier, 13 and 6 June 1648, A.G. A¹ 110, fols. 42, 14; Boissac to Le Tellier, 14 June 1648, from Casal Major, A.G. A¹ 110, fol. 45; Goury to Le Tellier, 9 June 1648, A.G. A¹ 110, fol. 23.

21. Regarding Falcombel, see Servien to Le Tellier, 6 June 1648, A.G. A¹ 110, fol. 14. For Jean Martin in Catalonia, see Goury to Le Tellier, 20 August 1648, A.G. A¹ 106, fol. 143. Goury's contract with Martin is mentioned in Champlastreux's instructions, A.G. A¹ 103, pp. 172r–84r. The correspondence of Clermont, 8 July 1648 from Dunkirk and 5 August 1648 from Calais, A.G. A¹ 106, fols. 105, 125, mentions munitioneer Fly.

André, *Michel Le Tellier et l'armée*, p. 441 with n. 1, maintained that the *conseil du roi* reserved for itself the granting of contracts to munitioneers after competitive bidding. André noted Goury as an exception; but I found other exceptions, for the royal council's control was not as complete in reality as André claimed.

accumulating large amounts of grain in the army magazines. As a result of the intendant's efforts, the French had stored some thirty thousand *quartiers* of oats in the fortresses of Barcelona, Santa Coloma de Queralt, Montblach, Balaguer, Flix, and Tortosa, worth some eighteen thousand *pistoles.* This presented problems, however, for the grain was not given gratuitously to the cavalry, but sold at low cost: one *quartier* for fourteen *reaux.* This amounted to a loss of ten thousand *livres* if the grain would have been sold on the open market. Goury therefore suggested that the crown might benefit if he had Le Tellier's permission to raise the price of grain sold to the native Catalan cavalry, although the price to the French would remain the same.[22]

Goury also concerned himself with the problem of logistics, the means of transporting supplies from France. This was not a problem for the army intendant alone, since he needed the cooperation of other officials inside France, especially the provincial intendants. Although it was easier to ship the supplies from French ports to Barcelona than to haul them across the Pyrenees, the various tolls and customs along the way posed difficulties, for France under the ancien régime had numerous internal customs barriers. These tolls, along with contractors' charges, boosted the cost of supplies to exorbitant heights, so the government issued special passports granting exemption from tolls to the entrepreneurs who carried supplies to the army.

This did not always work. A Monsieur Pestalosi, who had signed a contract with Goury to furnish twelve thousand *quartiers* of oats, ran afoul of the city council of Lyon, for that town, despite letters from the king, held Pestalosi accountable for customs on the Rhône River and delayed shipment of grain to Barcelona in 1648. Pestalosi was not the only munitioneer thus detained. François Bochart de Sarron Champigny, intendant of the region of Lyonnais, brought a similar problem to Le Tellier's attention that same year: a Monsieur Bidaud had had the same difficulties on the Rhône River at Arles. In the end, Champigny had to guarantee to pay the captains of the vessels for any tolls levied before they would begin the journey to Catalonia.[23]

Once the grain arrived at the mouth of the Rhône and the boats set sail for Barcelona, a new threat appeared—the Spanish. News reached Goury in 1648 that two enemy ships and two of their galleys lay waiting off Tarragona to prevent the passage of supplies to Barcelona. There was not much that the intendant could do except worry, but he wrote to Le Tellier: "Nevertheless, I have given a strict order that the

22. Goury to Le Tellier, 26 June and 3 August 1648, A.G. A¹ 106, fols. 80, 123.
23. Pestalosi to Le Tellier, 12 June 1648, A.G. A¹ 110, fol. 35; Champigny to Le Tellier, 5 and 12 June 1648, A.G. A¹ 110, fols. 13, 36.

vessels, brigantines, and frigates . . . pass at whatever cost, because if we lack, we are ruined." [24]

The danger was not over when the supplies arrived in Catalonia. They were stored in magazines until needed, then transported by convoy across the country. Naturally the Spanish hoped to prevent this transport by sorties. Goury feared for his lifeline from Ceruere, although the French had sent a contingent of troops to fortify the garrison. The Spanish also threatened the transport of supplies from the magazine at Flix, forcing Marshal Schomberg to go in person to secure that garrison. All these worries preoccupied the mind of Goury, and he could only shake his head and assert: "It is not that perhaps we will be unsuccessful in surmounting these difficulties, but for you [Le Tellier] to recognize that if the success is not all that we hope for, it will not be my fault, for I have indeed represented the inconveniences that can occur and have applied all the remedies possible to give the court the satisfaction it can desire in this enterprise." [25]

This dependence upon other French officials in matters of supply was not only a basic feature of the intendancy of the army in Catalonia, but also true for other armies. In 1647–48, when Mazarin planned a campaign in Italy, he needed such men as Alexandre de Sève, intendant of Provence, and Henri d'Estampes de Valençay to assist the army intendant in embarking troops for Italy. Sève had charge of the soldiers until they reached the port of Toulon; as the regiments marched through Provence under the direction of the *commissaires des guerres*, they fed on food stockpiled by the provincial intendant. Simon Arnauld d'Andilly, the future marquis de Pomponne, was intendant of the land forces attached to the naval expedition, and he awaited the arrival of the troops at Toulon. Valençay had the task of loading grain supplies on the ships, but Arnauld was to assist him, deciding how much grain should be converted into biscuits and placed on board. Arnauld and Valençay jointly negotiated the agreement with the captains of the vessels for the payment of transport and subsistence charges, although Arnauld's instructions called for him to pay no more than six *sols* per day, including bread at two *sols* per soldier. Arnauld also made a separate agreement to transport the horses and to feed them. Although Valençay, not Arnauld, was to pay the captains the sums agreed upon, the army intendant remained fully informed. Close cooperation between provincial officials and army intendants, therefore, was necessary for the launching of an army expedition. [26]

---

24. Goury to Le Tellier, 26 June 1648, A.G. A¹ 106, fol. 80.
25. Ibid.
26. Infreville to Le Tellier, 22 June 1648, from Toulon, A.G. A¹ 106, fol. 70. For the expenses of the soldiers on march to Toulon, see Champigny to Le Tel-

Armies sometimes needed officials to act in liaison with the local populace. There was no problem if the armies operated inside France, but complications arose when the armies operated on foreign soil. Army intendant René de Voyer d'Argenson solved his problem in Catalonia in the 1630s by becoming a virtual warlord in Barcelona, even dominating municipal affairs. But Argenson was an exceptional example. After his departure, the French used other agents in Catalonia: a native governor, Don Joseph Marguerita, and especially a French visitor general, Pierre de Marca, bishop of Conserans. Champlastreux's instructions mentioned earlier in this chapter urged him to have a close correspondence with Marca, "who in the absence of My Lord the Prince [Condé] will take care, with the sieur Don Joseph Marguerita, governor of Catalonia, of all that concerns the service of His Majesty in Barcelona and the assistance that can be desired from the said city and the country for the army." Marca's commission as visitor general granted him powers of inspection over all officers of justice and administration in the occupied territory, and he soon rose to be second in power to the viceroy. His authority made him the link between the French army and local officials in Barcelona.[27]

There was a different situation in Italy. From 1643 to 1648, the French fought in the peninsula as an ally of Savoy and other Italian princes against the might of Spain. An Italian, usually Prince Thomas of Savoy, shared command with a French general in a predominantly French army. One of the permanent intendants of this Franco-Savoyard army was Ennemond Servien, brother of the French minister Abel Servien. The king joined the ambassadorship of Savoy to Servien's intendancy, and with the combined powers of ambassador-intendant, Servien smoothed the path of the French army in Italy. He was particularly helpful in the matter of quartering portions of the French army in Savoy during the winter months. Thus in the autumn of 1648, the government sent a letter to Servien announcing the number of troops to stay in Savoy during the coming winter and reminding him to communicate the French desire to the duke of Savoy and his ministers "and to tender your offices to them, as you see convenient, for the lodging and subsistence of the said troops in the appropriate places and in the accustomed manner in Piedmont."[28]

lier, 23 June 1648, from Lyon, A.G. A¹ 110, fol. 67. Instructions given to Simon Arnauld d'Andilly exercising the charge of "intendant de la justice police et finances en son armée de terre qui sera jointe à son armée navalle," 12 May 1648, A.G. A¹ 107, pp. 240r–48v.

27. Sanabre, *La Acción de Francia en Cataluña*, pp. 148–50, 249–52. Champlastreux's instructions, A.G. A¹ 103, pp. 172r–84r.

28. The king to Servien, 29 November 1648, A.G. A¹ 108, pp. 194r–95v. Servien had considerable service in Italy. He was a "commissaire générale des guerres et

Although Servien was an ambassador, he was not the only army intendant to conduct negotiations with foreign governments. Jean Balthazar, intendant of the army in Lombardy commanded by the duke of Modena and Du Plessis Praslain, had recourse to missions to the courts of Modena and Parma although he was never an ambassador. A letter of Balthazar in October 1648 informed Le Tellier that the army commanders had sent him to the Parmese court to arrange the passage of the allied troops to Piedmont. At the same time, he hinted mysteriously that he had "entered into conference regarding some other circumstances which do not permit themselves to be committed to paper and which it is necessary to settle by word of mouth." Balthazar hoped to have a second audience later that day. Unfortunately, such tantalizing scraps merely hint at the diplomatic role that intendants of the army might perform when necessity demanded. No one has examined this problem closely, yet from the above evidence one may conclude that by 1648 the intendants had steadily expanded and consolidated their powers, compared with the army intendants under Richelieu.[29]

In one area, the army intendants' authority seems to have atrophied—the administration of justice. In comparison with their activity in the 1630s, intendants in the 1640s recounted only a few instances of their intervention in military justice. Servien once reported, with some relish: "Today I had executed three cavaliers [of the regiment] of Brauer who had robbed and killed two honorable merchants of this city [Turin] last year." But his example is almost unique. When an army intendant tried to administer justice in a foreign land, he discovered problems that a provincial intendant did not face. Thus differences arose with the government of Savoy, which complained that Piedmontese evaded native justice by enrolling in the French army, and demanded their extradition. Mazarin accepted this grievance and ordered Servien to hand over to local officials any Piedmontese who had committed a crime before his enlistment in the army. Moreover, in winter quarters in Savoy the intendant had the authority to handle

contrôleur des fortifications" at Pignerol in 1633, then "intendant de justice au delà des monts" and president of the sovereign council of Pignerol in 1645. In 1648 the crown appointed him intendant of the army of Italy and ambassador to Savoy. I did not find his actual commission, but there are many references to him as "intendant of the army" in the war archives at Vincennes. See the short biographical sketch of Servien in Charles-Prosper-Maurice Horric de Beaucaire, ed., *Savoie-Sardaigne et Mantoue*, Recueil des instructions données aux ambassadeurs et ministres de France depuis les traités de Westphalie jusqu'à la révolution française, vol. 14, pt. 1 (Paris, 1899), p. 3.

29. Balthazar to Le Tellier, 26 October 1648, A.G. A¹ 106, fol. 201. Although Balthazar promised, "J'espere vous mander dautres particulieres apres que j'auray conferé icy," I could find no further mention of this incident in the war archives.

crimes only among the soldiers; any complaint regarding soldiers and inhabitants was to be referred to the local authorities. But the intendant did retain his prerogative of exercising justice on campaign or in garrisons, even if among soldiers and inhabitants, no matter what the nationality.[30]

This policy tended to limit an intendant's judicial powers, and an examination of the correspondence between the years 1643 and 1648 indicates that army intendants were frequently requested to provide information about crimes in the army, but rarely did they act as judges. In 1647 when Nicolas Fouquet was intendant of the army in Flanders, the secretary of war asked him to investigate the circumstances surrounding a duel between the comte d'Harcourt, a brigadier, and the sieur de Vassé, *maître de camp* of the Piedmontese regiment. Fouquet's instructions required him to send this information "closed and sealed" to the king.[31]

Later that same year, Fouquet conducted another investigation—this time regarding the surrender of the fortress of Dixmude and the possible treason of its commanding officer, the sieur de Clanleu. Marshal Rantzau had taken the city from the Spanish on 13 July 1647 and given it a French garrison, but the fortunes of war shifted, and in October, Archduke Leopold approached the fortress. Although the commanding officer had assurances that the French would bring succor, he surrendered to the Spanish. Events revealed later that if he had held out for another six hours, the French would have arrived in time to save the garrison. The war department regarded the surrender as a serious matter, ordered the arrest of Clanleu, and requested Fouquet once

---

30. Servien to Le Tellier, 6 June 1648, A.G. A^1 110, fol. 14; Le Tellier to Servien, 29 May 1648, and to the marquis de Ville, same date, A.G. A^1 107, pp. 266r–67r, 264v–65v. As early as 1638, foreign troops in French service had difficulty recognizing the authority of the intendants of justice in the army "a cause qu'ils ont entr'eux des officiers proposez pour leur justice particulière." Since the crown admitted "qu'il est impossible que lesdictz. intendants puissent s'employer utilement a la police et discipline desdictes troupes a l'establissement et manutention des ordres qui pouront convenir leur subsistance, s'ils ny ont lauctorité requise," it ordered the officers of the foreign troops to recognize the intendant's authority. "Ordre aux troupes estrangers estants au service du Roy de reconnoître les intendants des armées de S.M.," 17 February 1638, from Saint-Germain-en-laye, A.G. A^1 49, fol. 50.

31. The king to Fouquet, 9 July 1647, A.G. A^1 104, p. 17rv. Intendants of provinces received similar orders from the crown; for example, on 18 July 1647, nine days after the letter to Fouquet, the government sent a letter to François Villemontée, intendant of justice, police, and finance "en province et pays de Poitou villes et gouvernment de la Rochelle et Brouage, Ile de Rhé et D'Olleron" (according to his commission dated March 1646, A.G. A^1 96, fol. 136), asking him to investigate the circumstances surrounding a duel between the sieurs des Rosches-Baritaud and de La Boullayée, A.G. A^1 104, p. 28rv.

again to uncover the true facts in the case and to send the information "closed and sealed" to Le Tellier. Fouquet played no other judicial role in the matter.[32]

Likewise, army intendants played no significant role in the most important military trial of the period. The French government tried Marshal de La Mothe-Houdancourt ostensibly for financial irregularities in Catalonia, but also for his military defeat at Lerida and his hostility to the queen and Mazarin. The crown recalled La Mothe-Houdancourt at the end of 1644, replaced him by the comte d'Harcourt, and arrested him on his return to France. His trial took place before the Parlement of Grenoble and dragged on until the Fronde secured his release. La Mothe-Houdancourt's army intendant, Michel d'Aligre de Saint-Lié, also was held in suspicion and ordered "to render account of the expenses of the past year," but there is no record of any action taken against him. The government did not confide the inquest of the La Mothe-Houdancourt affair to Aligre's replacement, but instead to Marca, the visitor general in Catalonia, and various provincial intendants. The new army intendant, Pierre Goury, seems to have played no role whatsoever in bringing the malefactors to justice. This was a far cry from the activity of intendants like Laffemas and Machaut in the 1630s.[33]

INTENDANTS OF CONQUERED TERRITORIES

Another type of intendant became common during the period 1643–48, the intendant of a newly conquered place. This type of intendancy was not an innovation; for example, in 1638 Henri Hurault de l'Hôpital, sieur de Bellebat, served as "intendant of justice, police and supplies in the lands of Upper and Lower Alsace." Due to the fortunes of war in the years of Le Tellier's secretaryship, this type of intendancy proliferated. France had fought continuously since the 1630s, and after the unfortunate year of Corbie in 1636, the tide of battle turned. In 1643 Condé's victory at Rocroi strengthened the Regency at home at the same time as it added to its prestige abroad. In the following years, French armies on the northern and eastern frontiers won an impressive number of towns: Thionville in 1643; Gravelines, Freiberg, Philippsburg, and Spires in 1644; Mardyck, Saint-Venant, and Armentières in

---

32. Chéruel, *Histoire de France pendant la minorité de Louis XIV*, 2:336, 424. The king to Fouquet, 30 October 1647, A.G. A¹ 104, pp. 210r–11r.

33. Louis André, "Le Maréchal de la Mothe-Houdancourt (son procès, sa rébellion, sa fin)," *Revue d'histoire moderne* 12 (1937): 5–35, 97–125, particularly pp. 97–99. For a short summary of the whole affair, see André, *Michel Le Tellier et l'armée*, pp. 321–26.

1645; Courtrai and Dunkirk in 1646. Although no peace had yet been signed, the French had to administer the newly conquered Flemish and German cities, which they hoped to make permanent possessions. For this task, the government resorted to a specialized type of intendant, that of the occupied territory.[34]

Unfortunately, little is known of these men except a few names and titles on assorted documents. In 1643 a Monsieur Dosny, "intendant of justice, police and finance in our cities and frontier places of our province of Champagne and that of Picardy," accompanied various other officials, including Jacques Dyel de Miromesnil, the regular provincial intendant of Champagne, to evaluate the domains of Sedan. Another agent, the sieur de Croisilles, received a commission in 1643 as "intendant of finance, fortifications, and supplies in the places on the frontier of Champagne." At the end of the following year, the government sent a commission to Charles Brèthe de Clermont; although not actually named an intendant by the document, he was to "have the intendance and direction" of the payment of troops and other expenses "concerning His Majesty's service in the lands of the Boulonnais and other reconquered areas." In 1646 an ordinance permitted the use of *revenants-bons*, surplus money, from the payment of the garrison at Fort de Watten, "according to the regulations of the sieur de Clermont, intendant of justice, police, and finance in our frontier places of Picardy and Artois, and in Flanders." In 1645 the war department dispatched a new commission to Dosny as "intendant of our finances and fortifications on our frontier of Champagne and in adjacent areas," for the construction of redoubts along the Meuse River, while Jean Colbert du Terron received a commission in 1647 to "have the intendance and direction of finances, contributions, supplies, and fortifications at Courtrai."

In that same year, 1647, Le Tellier wrote a letter to the receiver of domains at Hesdin, the sieur de Herly, asking him to render an account to René Le Vayer, "intendant of justice and finance at Arras," for some levies he had made without orders. Le Vayer's department included a number of conquered towns, for he had orders to furnish grain to the troops stationed at Arras, La Bassée, and Béthune. In 1648 the king named Le Vayer and Jacques de Chaulnes, both "councilors in our councils of state and privy," to take possession of the conquered towns

34. Commission of Bellebat, 20 April 1639, as "intendant de la justice, police, finances et vivres dudit pays de la Haute et Basse Alsace," A.G. A¹ 56, fol. 92; amplification of this commission to include the county of Montbéliard, 24 May 1639, A.G. A¹ 56, fol. 122 (both of these documents are also in Livet, *L'Intendance d'Alsace*, pp. 920–23). Chéruel, *Histoire de France pendant la minorité de Louis XIV*, 3:27–28.

in Artois ceded to France by the treaty of Münster, and Charles Brèthe de Clermont replaced Le Vayer as "president in chief of the sovereign council of Artois and intendant of justice, police, and finance in our *ville et cité* of Arras and of other towns and places in Artois."

When Clermont moved to Arras, he gave up his former position as "intendant of justice, police, and finance on the frontier of Picardy and in the newly conquered and in the advanced areas in Flanders along the seacoast," and Gamin replaced him. Gamin's commission accorded him the grandiose title of "intendant of justice, police, and finance in our frontier areas of our province of Picardy, even at Calais, Boulogne, Montreuil, Ardes, and other towns, places, and strongholds of the newly conquered lands, counties of Boulogne, Guignes, and Ardes, and even in the towns of Gravelines, Dunkirk, Bergues, Furnes, Bourbourg, and other areas newly conquered in Flanders toward the sea." There was at least one other frontier intendant in 1648, Louis Gombault, who was captured by the Spanish near Armentières. Little is known of him, but documents concerning his imprisonment titled him intendant of finance and fortifications in the lands occupied by the royal armies in Flanders.[35]

Although the intendants of newly conquered areas performed a large number of functions, undoubtedly the most important was that of levying contributions both in money and in goods upon the newly acquired populace. As the financial crisis deepened in France and with the military treasury almost empty, the crown tapped the occupied territories for funds, generally to pay and feed the forces of occupation. Yet the money was sometimes used for other purposes: in 1648 Le Tellier instructed Colbert du Terron to pay the baron de Palluau the sum of two thousand *livres* a month to support his expenses at

---

35. Commission of Miromesnil, Jeannin, Dosny, Boucherat, and Le Fèvre "pour l'evaluation du domains de Sedan," 4 January 1643, A.G. A¹ 80, fol. 1; commission of Croisilles, 12 June 1643, A.G. A¹ 79, 2d part under ministry of Le Tellier, fol. 24; commission of Clermont, 4 December 1644, A.G. A¹ 86, fol. 285; "Ordonnance pour employer les derniers revenants bons du payement de la garnison du fort de Waten," 10 May 1645, A.G. A¹ 92, fol. 255 (*revenants bons* was the surplus of money in an army after budgeted expenses had been paid); commission of Dosny, 18 April 1646, A.G. A¹ 96, fol. 141; commission of Colbert du Terron, 25 October 1647, A.G. A¹ 104, pp. 201v–5v; the king to sieur de Herly, 10 October 1647, A.G. A¹ 104, pp. 141v–43r; various letters to Le Vayer, 8 and 11 September 1647, A.G. A¹ 104, pp. 88v–89r, 90v–91r; commission of Chaulnes and Le Vayer, 26 October 1647, A.G. A¹ 104, pp. 205v–8v; commission of Clermont, 14 August 1648, A.G. A¹ 108, pp. 79v–83v; commission of Gamin, 12 November 1648, A.G. A¹ 108, pp. 165r–68r; Le Tellier to M. de Drone, concerning the imprisonment of Madame Gombault's husband, 1648, A.G. A¹ 110, fol. 232; see also the king to Marshal Rantzau, concerning demands for Gombault's liberty, 14 November 1648, A.G. A¹ 111, fol. 75.

Courtrai. Terron was to deduct this sum from the fund of contributions levied on Courtrai.[36]

The secretary of war sent numerous instructions to these frontier intendants regulating the contributions. When troops at Dixmude needed supplies for their winter quarters in that garrison, Le Tellier gave Clermont special permission to levy outside his department to amass the forage needed. In fact, Le Tellier's instructions sound like permission to raid: "impose and levy contributions in money, grain, oats, and forage in all the places of the said *franc* of Bruges where you can extend them" and bring the booty to Dixmude in the shortest possible time.[37]

Meanwhile, in Clermont's regular department the cavalry regiment of Chouppes, comprising eight companies, took up garrison for the winter. They needed to be paid two *môntres* in order to subsist, and Le Tellier's instructions were precise: Clermont had to calculate the amount of money needed to pay the two *môntres* to the eight companies on the basis of forty cavaliers per company, together with their officers. A captain was to receive 412 *livres* 10 *sols*, a lieutenant 272 *livres* 10 *sols*, a cornet 187 *livres* 10 *sols*, a *maréchal des logis* 145 *livres*, each cavalier of the light cavalry 40 *livres*, each of the six noncommissioned officers 30 *livres*, and 500 *livres* for the staff headquarters. After computing the sum needed, Clermont was to levy the amount upon all the districts in the Boulonnais, "as equitably as possible as to the resources of the area" and in concert with Villequier, governor of Boulogne. The intendant was to pay the two *môntres* in four lump sums, but only to the officers and cavaliers present and in an effective fighting force, and after deducting the cost of the supplies already furnished. Hay and straw for the horses had been given free, but the officers had to pay for their forage as well as for their own living expenses, except for the *ustensile*—the bed, linen, fuel, and candles given by the households in which they were quarterd.[38]

The practice of levying sums of money, oats, hay, and straw for the garrisons and troops in winter quarters became a regular institution along the northern frontier. Each summer the intendants of conquered areas busied themselves amassing vast quantities of forage, because the magazines depleted by the quartering of troops had to be replenished in the short summer months. These provisions fed not only the men in

36. The king to Colbert du Terron, 24 January 1648, A.G. A[1] 107, p. 36rv.
37. The king to Clermont, 13 September 1647, A.G. A[1] 104, p. 99rv.
38. The king to Clermont, 10 December 1647, A.G. A[1] 104, pp. 293v–95r. A *monstre* or *môntre* was a review of troops and, by extension, the wages paid to the soldiers after such reviews. Sometimes the *ustensile* was converted into a monetary payment too.

garrison or in winter quarter, but also any French troops that passed through the area. There was nothing unusual in this, as it was common practice elsewhere in France. Thus Le Tellier ordered Beaubourg, intendant of Lorraine, to impose contributions in money and in kind for the maintenance of the corps of soldiers under General d'Erlac, who had separated from the main army in the Empire under Turenne.[39]

These impositions on the conquered territories were not attempts to suck the areas dry. France hoped to retain these places after the peace and had no reason to destroy them. Then too, Le Tellier had a sense of justice, for he repeatedly asked the intendants to levy the contributions as equitably as possible, according to the resources of the region. In one letter to Clermont, Le Tellier asked the intendant not to allow any exemptions from the contributions, no matter what the type of rank or station. In another, the secretary permitted the communities around Arras and Béthune to deduct the cost of transporting grain to those garrisons from their normal contribution. Finally, Le Tellier desired that the contributions be levied only upon the government's orders; commissioners such as the sieur de Herly who levied sums without an order had to explain their actions to the intendant. In this manner, a rough justice was achieved in the matter of contributions.[40]

One might well pose the question whether the intendants obtained large sums of money from the contributions. Much of the frontier had seen continuous warfare for the last decade and a half; consequently it was hardly in a condition to pay much money. In 1648 Clermont was

39. The king to Clermont, concerning the amassing of forages for the cavalry in his department, 21 July 1647, A.G. A¹ 104, pp. 38v–40v; the king to Le Vayer, 11 September 1647, concerning the replacement of grain in the magazines of Flanders, A.G. A¹ 104, p. 91rv; the king to Beaubourg, 23 April 1648, A.G. A¹ 107, pp. 172r–73r.

40. The king to Clermont, 10 December 1647, A.G. A¹ 104, pp. 293v–95r: "vous fassiés l'imposition sur tous les lieux du Boulonnois . . . observant de le faire avec le plus dEgalite quil pourra en esgard a la force des lieux." The king to Clermont, 1 February 1648, A.G. A¹ 107, p. 50rv; the king to Le Vayer, 11 September 1647, A.G. A¹ 104, pp. 90v–91r: "vous imposiés et taxiés les villages sujets a contributions dans les pays d'Artois a fournir les charrois que vous verres leur estre possible et que vous leur diminuies sur leurs contribution en argent ce que montera les frais des voitures jusques a Bethune ou autre lieu quil conviendra porter des bleds et farines, a la raison que vous arbitraires et verres estre juste, et que vous fassies mention de cette diminution dans les mandements que vous envoyerés pour cet effet dans les villages." The king to the sieur de Herly, 10 October 1647, A.G. A¹ 104, pp. 141v–43r. Sometimes the crown exempted those who had given money for the transport of supplies from the quartering of troops; see the king to Clermont, 21 July 1647, A.G. A¹ 104, pp. 38v–40v, which reads in part: "Voulant que les contribuables qui satisfairont a la voiture des danrées et au payement des sommes auxquelles vous les aures taxés soient exemptes du logement de mes gens de guerre et soient au surplus soulagés et traittés le plus favorablemt quil se pourra."

pessimistic about his ability to extract money from his department. Conditions became worse when the French decided to withdraw troops up to the environs of Saint-Omer and Aire. Clermont remarked that if these areas were freed from paying contributions, he doubted if the rest of Flanders would pay, little as they actually gave. He even had the temerity to suggest that it might be more advantageous to give up the contributions altogether than to continue the expense of collecting them. As for the district of Furnes, there was no hope for that castellany, for the army had left it a desert. At the same time, Clermont thanked Le Tellier for giving him permission to deduct his back pay from the contributions, but suggested that the government seek other ways of paying him, for the army owed sixteen thousand *livres* in excess of the sum that might be expected from the contributions.[41]

Indeed, the contributions did not meet expenditure, and the army faced a chronic shortage of funds. With a feeling of fatigue, Clermont reminded the secretary once again that the sums advanced by Fly, the munitioneer, had not been repaid, although Fly had used up his own credit and consequently the army would lack bread. Meanwhile, Marshal Rantzau continued to demand all sorts of things from the intendant, including lumber and three hundred muskets costing a total of five thousand *livres*. The financial situation was desperate, Clermont claimed: "I have shown you what our contributions equaled by the statement I sent to you, not including the 1,900 *livres* which I reimbursed Madame de Rantzau for the money she lent for the barracks of the cavalry . . . nor a thousand *livres* of other expenses." Where the money would come from, Clermont did not know. He awaited Le Tellier's instructions.[42]

The intendants of newly conquered areas had other responsibilities in addition to supervising the contributions. In one sense they were much like provincial intendants, for they had jurisdiction over a specific department. But one must not think of these departments as *généralités* or provinces with traditional boundaries, for the word *département* had a much looser connotation, merely meaning the area over which one exercised one's commission. Administrative decrees from the government regulated the boundaries of these areas, and in the case of

41. Clermont to Le Tellier, 8 July 1648, A.G. A¹ 106, fol. 105. Regarding the payment of the intendant's wages, see the remarks of Clermont to Le Tellier, 5 August 1648, A.G. A¹ 106, fol. 125: "l'honneur quil vous a pleu me faire de me donner assignation sur ce fond pour cinq mil livres des appointemens extraordinaires qui me sont deubt, n'empeschera pas s'il vous plaist que ie ne vous importune a vous demander un autre moyen de men faire payer puisque nous devons XVIg# plus que le fonds, si ie n'estois asseuré de vostre bonté par tant de bienfaits, que iay receu de vous, ie ne serais pas si hardy q. d'en user ainsy avec vous.
42. Clermont to Le Tellier, 5 August 1648, A.G. A¹ 106, fol. 125.

newly conquered territories in Flanders, the government had no hesitation in shifting areas from the control of one intendant to another.

Thus on 7 February 1648 Le Tellier wrote letters to Clermont, Le
Vayer, and Colbert du Terron dividing the various dependencies paying contributions, so that there would be no confusion about overlapping jurisdictions. Evidently this arrangement did not prove satisfactory, for the following April the government wrote to René Le Vayer,
intendant at Arras, who had the town of La Bassée under his jurisdiction, and to Colbert du Terron, intendant of finance and contributions
at Courtrai, defining the areas in which each would levy contributions.
Previously, everything beyond the Deûle River from Warneton to
Lille and from Lille to Pont-Tessin was subject to La Bassée, and everything beyond contributed to Courtrai. But the villages of Hellemmes,
Flers, Orcq, and Hanappes had traditionally contributed to Courtrai,
because they were farther from La Bassée than from Courtrai and the
parties from La Bassée could collect contributions from these areas only
with difficulty. Therefore the government redrew each jurisdiction.
The villages of Hellemes, Flers, Orcq, Hanappes, Chin, Grugen, Cisoing,
Cobrien, Ginen, Romps, and all other towns along the route from
Pont-Tessin and Orchies, and everything from Orchies to Marchiennes
going toward the Escaut River to Saint-Amand, and from Saint-Amand
along the Escaut to Tournai, would pay contributions to Courtrai,
leaving the other areas to La Bassée.[43]

The intendants of occupied areas were the royal agents of administration in their departments. Like the provincial intendants inside France,
every aspect of territorial life came under their jurisdiction. They
even supervised the election of city magistrates. Marshal Rantzau and
Clermont, for example, received instructions in 1648 to conduct the
municipal election of the officials of Bergues–Saint Vinox. On another
occasion, Le Tellier sent Clermont instructions for the election of a
new abbot of Auxi: assemble the monks of the abbey under the supervision of the abbot of Saint-Vaast and proceed to a new election. In
another letter, Le Tellier once again interfered in local religious affairs,
forbidding the furnishing of army rations to various religious houses
along the frontier and urging intendants Le Vayer, Dosny, Gamin, and
Colbert du Terron to correct this abuse. Thus it is apparent that an
intendant's jurisdiction extended to everything, no matter how petty

43. The king to Clermont, with a copy of the same to Le Vayer and Colbert du
Terron, 7 February 1648, A.G. A¹ 107, pp. 58r–59r; the king to Le Vayer, with
copies sent to Colbert du Terron and *commissaire des guerres* Jacques Charuel,
10 April 1648, A.G. A¹ 107, pp. 155v–56v. I have modernized the spelling of towns
in the letters insofar as I could discover them on modern maps. In the original,
Lille is spelled "L'Isle," Warneton "Varnettan," and so forth.

it seemed. In this manner, royal authority entrenched itself in the newly conquered territories.[44]

From the above examples, one might easily conclude that such intendants more closely resembled regular provincial intendants than intendants of the army. The solution is not that simple, however, for in many of their actions these frontier intendants could pass as army intendants. Various correspondence between the secretary of war and the intendants in 1648 illustrates this situation. On 30 June 1648 Le Tellier sent an ordinance regulating the division of the spoils of war between the cavalry and infantry in Champagne, leaving the enforcement to the prince of Conty, and in his absence to a lieutenant general in the area and to Dosny, intendant of finance on the frontier. Earlier that spring Le Tellier had written to Marshal de La Meilleraye, Lieutenant general of the army of Flanders, approving the payment of bonuses to captains of infantry whose troops looked good in the reviews, as well as the long-overdue back pay for winter quarters. Although the sums were to be paid upon the marshal's orders, they were to be countersigned by intendant Le Vayer. Such financial arrangements were similar to those of other army intendants in Catalonia and Italy. Likewise, Le Tellier demanded that the intendants of occupied areas submit the same reports and statistics as did the other army intendants, and with the money from the contributions providing the support of the army, it is not difficult to see the reason for Le Tellier's desire for an accounting to the war department.[45]

Another facet of an army intendant's wide range of duties was his supervision of fortifications. For example, Goury in the army in Catalonia had charge of the fortresses in that area, allotting funds, supervising construction, and contracting for supplies. In this he resembled an intendant of a newly conquered place, who had similar responsibilities. Dosny, "intendant of our finances and fortifications on our frontier of Champagne and areas adjacent," had orders to construct fortified places along the Meuse River, with authority to levy a portion of the sums

44. Commission of Marshal Rantzau and Clermont to proceed with the municipal election of Bergues-Saint Vinox, 1 March 1648, A.G. A¹ 107, pp. 92v-94r; the king to Le Vayer, concerning the election of the abbot of Auxi, A.G. A¹ 104, pp. 152v-53r; ordinance of the king revoking the furnishing of bread to religious houses, 30 November 1647, A.G. A¹ 104, pp. 258v-6or.

45. Ordinance regulating the division of "butin," 30 June 1648, with accompanying letter to Dosny, same date, A.G. A¹ 107, pp. 313r-14v; the king to La Meilleraye, 6 April 1648, A.G. A¹ 107, pp. 152r-53v. Cf. also Clermont to Le Tellier, 24 June 1648, A.G. A¹ 106, fol. 74: "Je n'ay point encore les comptes de nos contributions pour vous en informer iay prié Mʳ deschamps de dresser le sien, que ie vous envoyeray aussitost affin que vous en disposiés, soit pour le rembourcement de Mʳ Fly, ou autres choses, quil vous plaira."

needed from the local populace. During his intendancy he supervised
the work of the engineer, Roses, and oversaw the construction of at
least twelve redoubts, each containing ten men.[46]

Intendants of conquered regions regularly visited the armies and
worked closely with the generals in matters relating to military life.
Clermont, for example, not only worked to feed and pay the armies,
but also assisted the arrival of new companies of foreign mercenaries
in the army. In July 1648 Clermont received sixty-nine English recruits
in Calais, the following month he welcomed the arrival of another fifty
Englishmen, and he expected another five hundred mercenaries in a
few days. Clermont did not think that Englishmen had the stamina to
last as soldiers, but he obeyed orders and equipped them. He was restricted, however, by the shortage of funds, and once again he asked
for Le Tellier's assistance "for these sorts of expenses." [47]

From his advance post on the frontier, Clermont, like ordinary army
intendants, sent Le Tellier military intelligence. From the vantage point
of Calais, he reported news of Rantzau's attempt on Ostend in June
1648: the successful disembarkment of the men from their boats, then
the violent sea that prevented the unloading of supplies and munitions,
thus isolating the men on the beach, where they were at the mercy of
the enemy. Clermont tried to get the news to Le Tellier as quickly as
possible, but the Spanish captured the messenger he dispatched to
France. Nevertheless, Clermont continued to report the activities of the
enemy, particularly rumors that the Spanish were buying ships from
the Dutch to attack Calais. Some reports said that the enemy had obtained thirteen vessels, while merchants with correspondents at Ostend
claimed that as many as twenty ships had set sail for the coasts off
Calais. Despite the fragmentary nature of the correspondence, it is obvious that, like army intendants, intendants of conquered places were
expected to inform the secretary of war of everything that happened
in their vicinity.[48]

One intendant in particular, René Le Vayer, exemplified the hazy
borderline between an army intendant and that of a newly conquered
area. In April 1647 Le Vayer was an intendant at Arras; one month
later Le Tellier named him to exercise the charge of intendant of jus-

46. Gaston Zeller, *L'Organisation défensive des frontières du nord et de l'est au
XVIIe siècle* (Paris, 1928), pp. 43–44. Dosny's commission to work on the redoubts, A.G. A¹ 96, fol. 141. Compare Goury's actions in Catalonia; see his letters
to Le Tellier, 21 June and, more important, 3 August 1648, A.G. A¹ 106, fols. 67,
123. Note that there were also engineers called intendants of fortifications, but
they were not intendants of conquered places.

47. Clermont to Le Tellier, 8 July and 5 August 1648, A.G. A¹ 106, fols. 105, 125.

48. Clermont to Le Tellier, 24 June and 5 August 1648, A.G. A¹ 106, fols. 74,
125.

tice, police, and finance in the army of Flanders until the arrival of the new army intendant, Fouquet, who had been delayed. According to his instructions, Le Vayer was to exercise "the same powers to which the intendants of my armies are accustomed by virtue of the commission that you have for Arras, without having need of any other expedition from me." Upon Fouquet's arrival, Le Vayer was to return to Arras without waiting for other orders from the king. It is significant that Le Vayer did not need a new commission to empower him to function as army intendant; nor did he need another document authorizing him to resume his intendancy in Arras. He simply passed from an intendancy of an occupied region to an army intendancy and back to an intendancy of an occupied region without any commotion.[49]

By the time of the Fronde, the institution of the military intendancies had acquired a maturity and expertise that would stand the government in good stead in the troubled years ahead. The office of intendant remained flexible—one witnesses the intendants of newly conquered provinces, standing halfway between army intendants and regular provincial intendants and combining the features of both. But although the intendants of occupied places were important in the Thirty Years' War, it was much later, after the monarchy had triumphed over its internal enemies, that the government once more extended its conquests. Before this could occur, the government faced a powerful threat from unruly subjects who resented the crown's increasing power at the expense of their own privileges. This event, the Fronde, menaced not only the institution of the intendants, but also the very integrity of the monarchy itself. Fortunately for the crown, it had the institution of the army intendants, one of the instruments it needed to restore order.

49. The king to Le Vayer, 23 May 1647, A.G. A¹ 103, p. 354rv. The relevant portion of the letter reads: "que vous pouvies sans faire prejudice a mon service dans Arras vous employer pendant quelque tems a cette intendance en mad. armee et jay bien voulu vous faire cette lettre pour vous dire qu'aussitot que vous l'aures receüe vous ayes a vous y rendre pour faire les fonctions d'intendant de la justice, police, et finances en icelle tant ainsy et avec les memes pouvoirs qu'ont accoutume de faire les intendans en mes armées et en vertu de la commission que vous en aves pour Arras sans que pour ce il vous soit besoin d'autre expedition de moy et lorsque led. Fouquet sera arrivé en mad. armée vous pourres retourner a Arras et lorsque vous le pourres faire avec seureté sans attendre autre ordre de moy." It was also customary for a new army intendant to consult with his predecessor on any information he would need to know in his new employ; thus there was a continuity of experience in the intendancy. For example, see the king to Imbert, announcing the arrival of Goury as his replacement in the army of Catalonia, 24 December 1647, A.G. A¹ 104, pp. 317r–18r, and informing him: "que vous luy en remettre les fonctions avec une connoissance entière des despences qui ont este faites sur les fonds destinés pour la Catalogne tant pour lentretenement des troupes que pour les fortifications de Catalogne et quen cette occasion et en toutes autres vous tenies une parfaite correspondance avec luy."

V

The Fronde and Its Aftermath
1648-59

The reign of Louis XIII and Richelieu had witnessed a dramatic increase in royal authority, a "Governmental Revolution," resulting from three key factors. First, warfare. France had clandestinely aided German princes in their struggle against the Habsburgs in the Thirty Years' War, then openly participated in the final phase of the struggle from 1635 to 1648. Even with the Peace of Westphalia, the crown continued its fight against Spanish encirclement until 1659. These years of perpetual fighting meant high costs, not only due to large and better-equipped armies, but also due to the years of campaigning. Moreover, the war effort, never popular with the French, necessitated rapid decisions and execution rather than long deliberations among traditional government institutions with their protection of local rights and privileges.

Second, economic problems. The first half of the seventeenth century saw a reversal of the preceding century's upward economic trend. This "depression," whether measured by demographic decline, bad harvests, epidemics, stagnation of prices, regression in land cultivation, decline in profits, or a slowdown in investments, meant that the government had a weaker economic base upon which to draw for its costly wars. Third, popular uprisings. Almost as a chain reaction to the government's attempts to squeeze more money out of the populace despite its depressed economic condition, a number of popular revolts flared. These uprisings—in which local aristocrats and towns often lent tacit if not open support—usually coincided with the arrival of the tax collector to collect some new levy.[1]

1. A. Lloyd Moote develops the theme of the "Governmental Revolution" in his recent book, but other French historians have used the term before; see Moote's remarks in *The Revolt of the Judges: The Parlement of Paris and the*

To meet these threats to its authority, the crown resorted to methods that could only alienate its subjects, including part of its own bureaucracy. Complaining of inefficient local officials—their slowness and corruption—the government turned to special commissioners both in the provinces and in the army to circumvent the normal but cumbersome judicial and fiscal routines. These special agents threatened the older officeholders, particularly the *trésoriers de France* and the *élus* who apportioned and levied the taxes. To meet its costs, the government increasingly resorted to *partisans* or *traittants*, tax farmers and financiers who charged high rates of interest but who would furnish the money the government needed. In order to protect their investment, these financiers insisted that the government guarantee tax collection by the use of force.

Provincial intendants were the key officials in this process. Bypassing the traditional treasurers, they apportioned taxes among the towns and parishes, they levied the taxes—often accompanying royal troops to aid in collection—and they sometimes created special taxes on their own authority that were later validated by decrees from the royal council. The *trésoriers* and *élus* were left with only technical collaboration and judicial formalities, such as registering commissions for the levy of the taxes. These officials saw their power and prestige whittled away by the intendants; and just as significant, they suffered economic loss from the decreased sale price of their offices and a decline in the customary remunerations—*droits*—that accompanied an office. Naturally they protested vigorously through their own corporate organizations, called *syndicats*, to other judicial and financial officials.[2]

Fronde, 1643–1652 (Princeton, N.J., 1971), p. 43n19. The rest of this paragraph is a summary of Roland Mousnier's perceptive remarks in "La Participation des gouvernés à l'activité des gouvernants dans la France du XVIIe et XVIIIe siècles," *Schweizer Beitrage zu Allgemeinen Geschichte* 20 (1962–63): 216; Mousnier has expanded on these themes in numerous articles conveniently collected into one volume under the title *La Plume, la faucille et le marteau* (Paris, 1970). Concerning popular uprisings, see his *Fureurs paysannes* (Paris, 1967), translated into English by Brian Pearce under the title *Peasant Uprisings in Seventeenth-Century France, Russia, and China* (New York and Evanston, 1970); see particularly pp. 322–25 on the support the privileged classes lent the rebels. For a succinct summary of the military changes taking place in the seventeenth century that necessitated the higher costs, see Michael Roberts's inaugural lecture delivered before Queen's University of Belfast, *The Military Revolution, 1560–1660* (Belfast, [1965]), reprinted in his *Essays in Swedish History* (Minneapolis, Minn., 1967), pp. 195–225. His remarks have recently been confirmed by Geoffrey Parker, *The Army of Flanders and the Spanish Road, 1567–1659*, Cambridge Studies in Early Modern History, no. 3 (Cambridge, 1972).

2. Roland Mousnier, "Recherches sur les Syndicats d'*Officers* pendant la Fronde: Trésoriers généraux de France et élus dans la Révolution," *XVIIe siècle*, nos. 42–

The sovereign courts became intimately involved in this struggle, not only because of ties of blood and friendship with these office-holders, but also because these were the courts that normally handled fiscal appeals. The most important of these courts, the Parlement of Paris, had particular reasons for hating the *traittants*. In many ways the judges represented a closed caste and resented attempts by the finan-ciers, with their immense wealth and often humble family origins, to infiltrate the prestigious parlement. Since this parlement had a long tradition of verifying new tax measures, it was certain to play an im-portant role in any tax dispute. In brief, the Parlement of Paris soon became the leader in opposition to the crown.[3]

The consequence was what one historian recently has called a "logical progression" from the first signs of parlementary resistance be-ginning in 1643 over clashes in finances and emerging in 1648 with a clear challenge to the crown. The smoldering tinder of the Fronde blazed into the open with the Parlement of Paris's assent to the Decree of Union, 13 June 1648. This decree had begun as one maneuver in the quarrel between the central government and the courts. After threatening not to renew the *paulette*, the annual tax that guaranteed an officeholder's proprietorship and the privilege of transmission of office, the government made a partial concession. Although the crown had granted the renewal gratuitously to the Parlement of Paris, the members of the other sovereign courts, including the Parisian *grand conseil*, the *chambre des comptes*, and the *cours des aides*, and the provincial parlements had to pay a sum equivalent to four years of their salary in order to obtain this privilege. The Parlement of Paris saw this as an attempt to divide the sovereign courts and to gain revenge for their obstructionism, so the Parlement of Paris agreed to unite in common assembly with the other sovereign courts in the Chambre Saint Louis. The government's clumsy attempt to suppress this assembly moved Parlement beyond a defense of its own interests to the broader issue of public reform, in an attempt to capitalize upon the widespread discontent in the kingdom. The first phase of the revolt, the Parlementary Fronde, had begun.[4]

43 (1959), pp. 78–84, and "Etat et commissaire: Recherches sur la création des intendants des provinces (1634–1648)," *Forschungen zu Staat und Verfassung: Festgabe für Fritz Hartung*, ed. Richard Dietrich and Gerhard Oestreich (Berlin, 1958), pp. 329–30, 340–42. For an example of such a provincial intendant, see Francis Loirette, "Un Intendant de Guyenne avant la Fronde: Jean de Lauson (1641–1648)," *Bulletin*, Comité des travaux historiques et scientifiques, section de philologie et d'histoire, 1957, pp. 433–61, especially pp. 447–55.

3. Mme. Cubells, "Le Parlement de Paris pendant la Fronde," *XVIIe siècle*, no. 35 (1957), pp. 178–82.

4. Moote, *Revolt of the Judges*, p. 91. Moote's chap. 4, pp. 91–124, is the most recent attempt to describe the complicated politics from the rise of the Parlement

The Chambre Saint Louis began its first session on 30 June 1648. Significantly, one of the first resolutions agreed upon was an article suppressing the office of intendant: "The intendants of justice and all other holders of special commissions not verified in the sovereign courts will be revoked from this time forward." Later the assembly reinforced this resolution with another, declaring that all ordinances and judgments rendered by the intendants of justice were broken and henceforth void, and forbade any subject of the king to appear before the said intendants upon pain of a ten thousand *livre* fine. Finally, so that everyone would know of the abolition, the assembly ordered its decree to be read in all corners of the kingdom.[5]

Faced with the assembly's strong opposition and cognizant of its own weakness, the government decided to compromise. On 6 July 1648 the advocate general, Omer Talon, reported to the Parlement of Paris the crown's willingness to agree that the commissions of the intendants, unless verified by the sovereign courts, were illegal, but pleaded that war precluded widespread reforms. The intendants were necessary for the war effort, the crown maintained, because only they could ensure efficiency and promptness in tax collection. "There is a great difference," the queen reminded Parlement, "between the employ of thirty-five persons who are established throughout the kingdom to keep order in the levy of royal taxes—who are the intendants—and that of three thousand persons, because the number of *trésoriers de France* and *élus* reaches this number."

If the collection of taxes were taken away from the intendants and given back to the regular officials, France would face economic chaos and failure in that year's military campaign. This did not satisfy

of Paris's opposition up to the Chambre Saint Louis. See also the older account of Paul Rice Doolin, *The Fronde* (Cambridge, Mass., 1935), pp. 6–10, although Ernst H. Kossmann, *La Fronde* (Leiden, 1954), pp. 34–53, is the most succinct account. Kossmann saw the Fronde from a sympathetic parlementarian viewpoint, that of a parlement pushed into revolution against its wishes and despite its belief in royal absolutism. Consequently Kossmann did not see the Chambre Saint Louis as inaugurating a revolution (p. 53), because it had no power except to counsel or suggest. Only when Mazarin tried to reduce his opponents by force during the siege of Paris in the opening months of 1649 did the Fronde actually begin. Moote, *Revolt of the Judges*, p. 144, is much more traditional when he suggests that the Chambre Saint Louis really began the Fronde: "In a literal, if incomplete sense, then, the revolt of the judges in the summer of 1648 can be called the 'parlementary Fronde.'" It should be noted that Roland Mousnier took a distinctly hostile view of the Fronde and what he called its "demagoguery" in his "Quelques Raisons de la Fronde: Les Causes des journées révolutionnaires parisiennes de 1648," *XVIIe siècle*, nos. 2–3 (1949), pp. 33–78.

5. The text of these resolutions is in the anonymous *Journal contenant tout ce qui s'est faict et passé en la cour de Parlement de Paris, toutes les chambres assemblées, sur le suiet des affaires du temps present* (Paris, 1652), pp. 12, 16 (hereafter cited as *Journal de Parlement*).

Parlement, which suggested the army could be paid for by forcing the financiers to disgorge their unjust and immense profits. Two days later, 8 July, parlementary leaders forced Mazarin—"harassed, excited, and annoyed," according to an eyewitness—to recall all intendants. Consequently on the following Saturday, 11 July, Talon formally announced the king's declaration revoking the commissions of the intendants of justice, police, and finance in the provinces under the jurisdiction of the Parlement of Paris. But there were exceptions, namely the intendants of the provinces of Champagne, Picardy, and Lyonnais, where "because of the frequent lodging and passage of troops, it was necessary to provide promptly for them." These three intendants, however, would have no authority over disputed judicial matters, except in cases between soldiers and civilians, nor any direction of finances.[6]

Once again these concessions did not satisfy the parlementarians. They believed that the government was stalling, since not all the intendants' commissions were revoked but only those within the territory subject to the Parlement of Paris. Even then there were exceptions such as in Champagne, Picardy, and Lyonnais; and in the excepted intendancies, the government made no mention of the need to register the intendants' commissions in the sovereign courts, thus removing a powerful bargaining position vis-à-vis the intendants.

Cleverly coupling its demands with a request that a special chamber of justice, composed of members of the sovereign courts, be established to investigate reports of illegal tax collections and financial corruption in the government, Parlement once again forced the crown to make partial concessions. On Thursday, 16 July, the duke of Orléans brought to Parlement the queen's agreement to abolish the intendants; in fact, letters had already been sent recalling them. But the crown would agree only to such reforms as it thought necessary to calm the tense situation, and no more. Thus the formal declaration abolished all provincial intendants except for six in frontier provinces: Languedoc, Burgundy, Provence, Lyonnais, Picardy, and Champagne. Although the decree doubled the number of reserved intendancies from three to six, the government had made concessions. It agreed that the six intendants would have nothing to do with the imposition and collection of taxes, but would serve "near to governors to assist them in the execution of their powers." Their duty was the organization of supplies to assist the war effort. Parlement reluctantly agreed to this declara-

6. Omer Talon, *Mémoires*, ed. Michaud and Poujoulat, Nouvelle collection des mémoires relatifs à l'histoire de France, vol. 30 (Paris, 1857), pp. 244–49; *Journal de Parlement*, pp. 30–31.

tion, yet specified that the commissions of the intendants must be registered in Parlement.[7]

The reform that the Parlement of Paris had forced upon an unwilling government was an attack on the provincial intendants, not on the army intendants or the intendants of newly conquered places. It was a protest against the fiscal powers of the provincial intendant and the supersession of the powers of the traditional tax officials, the treasurers and *élus*. It was also a tax revolt against the hated financiers and tax farmers, because—as one chief official of the Parlement of Paris, Jacques Le Coigneux, remarked—the people considered the intendants to be *valets des partisans* and not the king's men. Another president of Parlement, Nicolas Potier de Novion, agreed that the provincial intendants were the partners of the *traittants*, self-interested in the very matters they were sent to judge and enriching themselves at the expense of the countryside.[8]

Parlement had no quarrel with the army intendants. It had listened to the queen's remarks on 6 July 1648, when she painted a somber picture of the present situation: Condé, on the northern frontier,

7. *Journal de Parlement*, pp. 37–39. The declaration sent to the Parlement of Paris and registered there, 18 July 1648, is in the Michaud and Poujoulat edition of Talon's *Mémoires*, pp. 250–52. It is exactly the same as one sent to the Parlement of Grenoble, 12 July 1648, a copy of which is preserved in A.G. A¹ 108, pp. 29r–33v. A copy of a letter sent to François Villemontée recalling him as intendant of Poitou, 15 July 1648, is in A.G. A¹ 108, p. 12rv; it simply states: "Monsr de Villemontée, ayant pour diverses considerations resolu de rapeller les intendans de la justice et finances dans la plus part des provinces et generalités de mon royaume, je vous faits cette lettre par l'avis de la reyne regente madame ma mere que mon intention est que vous vous rendis au plutot pres de moy ou je seray bien aise de vous temoigner comme je suis entierement satisfait de vos services et que s'il s'offre occasion de vous en reconnoistre je le seray priant dieu quil vous aie Monsr de la Villemontée en sa ste. regard. Escrit a Paris le XV juillet 1648." As secretary of state, Le Tellier sent similar letters to Jacques de Chaulnes, intendant of Limousin, and Denis de Heere, intendant in Dauphiné.

Not all parlementarians believed that the reserved intendants were necessary for the military situation. The arch-Frondeur Pierre Broussel, for example, remarked: "que ceux que l'on choisissoit tous les jours pour cette sorte de commissions, n'estoyent le plus souvent que de jeunes gens qui ne sçavoyent ny finances, ny justice, ny police, et moins encore les ordres de la guerre pour lesquels on pretendoit les reserver dans les trois provinces." *Histoire du temps ou le véritable récit de ce qui s'est passé dans le Parlement depuis le mois d'aoust 1647 jusques au mois de novembre 1648. Avec les harangues et les advis différents, qui ont esté proposez dans les affaires qu'on y a solemnellement traittées* ([Paris], 1649), p. 133 (hereafter cited as *Histoire du temps*). There are various copies of this work with different pagination; I used the copy in the Newberry Library, Chicago.

8. Talon, *Mémoires*, p. 247. Cf. the remarks of Broussel in *Histoire du temps*, p. 132: "Monsieur de Broussel dit, que le nom d'intendant estoit si odieux et si suspect au peuple qu'il falloit en abolir et en oster la memoire, et ainsi que l'on n'en pouvoit reserver dans aucune province sans continuer les desordres ausquels on avoit voulu pouvoir."

needed bread and money for his soldiers, and if the army collapsed for want of supplies, the queen feared an enemy incursion into Picardy reminiscent of the "year of Corbie" in 1636. If the Swedish mercenaries did not receive their pay, they would treat with the enemy. The armies of Turenne, the landgrave of Hesse, and that in Catalonia were dependent upon money for the movement of supplies and the support of the artillery. French patriotism had led the Parlement of Paris to permit the existence of intendants in the six frontier provinces for the war effort—provided that "they have no disputed jurisdiction with the traditional officeholders, nor knowledge of finances, but only of the men at war." Parlement could do no less for the army intendants.[9]

This fact more than anything else illustrates the clear distinction between provincial and army intendants by the end of the 1640s. While the intendants of the provinces had become a permanent institution bringing about a revolution in administration inside the country, the army intendant's tenure was renewed each campaign year. Being outside France, his commission was beyond the territorial jurisdiction of the parlements and did not need registration. Furthermore, his powers did not alienate the corporate interests of the powerful officeholding class. Therefore it is not surprising that there was no mention of the intendants of the army in the records of the parlementary debates, even though these agents fell into the same category as those with *commissions extraordinaires*, which the parlements heartily condemned.[10]

It remained to be seen how each side would observe the agreement regarding the intendants. The government could not reconcile itself to the loss of its agents; indeed, as one historian has remarked, the military and fiscal demands of civil and foreign war made violation of the reform inevitable. Throughout the period of the Fronde, the government sent officials into the provinces as crypto-intendants, although they did not bear the title of intendant of justice, police, and finance. Some of these personnel were officeholders with special commissions or masters of requests on circuit, but the two most common substitutes were intendants of finance and army intendants. This infuriated the Frondeurs, for it was obvious that these substitutes would resume the

---

9. Talon, *Mémoires*, pp. 245–46. Cf. *Histoire du temps*, p. 122. Perhaps another reason for permitting the existence of certain frontier intendants was the fact that Jean-Edouard Molé de Champlastreux, son of Mathieu Molé, first president of the Parlement of Paris, was the intendant in Champagne; Moote, *Revolt of the Judges*, p. 175.

10. The abolition of the "commissions extraordinaires" is found in article 10 of the resolutions passed by the Chambre Saint Louis, in *Journal de Parlement*, p. 16. Cf. Moote, *Revolt of the Judges*, p. 160.

collection of taxes and continue to arrogate the role of the traditional fiscal officials.[11]

At first the government attempted to use members of the sovereign courts, such as judges of the *cours des aides* and parlements, although the response of the Parlement of Paris quickly ended this practice among its own members. Jacques de Boisson, seigneur d'Auxonne and first president of the *cour des aides* at Cahors, is an example of such a special commissioner. In 1651 the crown ordered him to Rodez to examine the accounts of the tax receivers in that area. The local officials in the bureau of finance at Montauban strongly protested this action as a usurpation of their own authority, and the Parlement of Toulouse ordered him to register his commission with that court under threat of imprisonment—undoubtedly a humiliation for a member of a sovereign court. Lack of documentation, however, prevents learning the result of this clash.[12]

The crown also resorted to dusting off the old ordinances calling for masters of requests to make regular inspection tours into the provinces—*chevauchées*—which had fallen into disuse with the emergence of the intendants. In January 1650 the government attempted to turn the clock back by sending five masters of requests with identical powers into five different provinces. One of these men was Thomas Morant, who received a *lettre de cachet*—not a regular commission, which needed to be verified in the courts—authorizing him to make an inspection trip into Guyenne. He was to "visit all the cities and places therein as you judge appropriate, inquiring into and preventing all disorders and malversations that have been committed or that will be committed concerning my finances and contraventions to my ordinances," punishing offenders, and to "generally exercise the functions attributed by the laws to the masters of requests . . . for the good of my service and the support of my subjects."

Upon Morant's arrival in Guyenne, the *trésoriers de France* complained to the Parlement of Toulouse; the parlement, in turn, forbade him "under whatever name and pretext that it be, to perform any function of an intendant." Although Morant attempted to conciliate the parlement, he was unsuccessful. Parlement ordered him from the city within twenty-four hours and prohibited any official from obey-

---

11. Moote, *Revolt of the Judges*, p. 259. Edmund Esmonin, "La Suppression des intendants pendant la Fronde et leur rétablissement," originally in the *Bulletin* of the Société d'histoire moderne, 1935, reprinted in his collected essays, *Etudes sur la France des XVIIe et XVIIIe siècles* (Paris, 1964), p. 34.

12. Jean-Paul Charmeil, *Les Trésoriers de France à l'époque de la Fronde: Contribution à l'histoire de l'administration financière sous l'ancien régime* (Paris, 1964), p. 390.

ing his orders. He retreated to await the arrival of a decree from the royal council confirming his powers, but the parlement accused him of disturbing the peace (*du repos public*) and sent its own officials to arrest him when he attempted to try a Spanish spy. This attempt failed, but Morant once more embroiled himself with the parlement when he tried to preside over the trial of a counterfeiter. The outcome of this is not known, but shortly thereafter Morant retired from the province.[13]

The crown also attempted to use fiscal agents, the intendants of finance, to substitute for provincial intendants. These officials were members of the accounting branch—the *contrôle général des finances* —of the department of finance headed by the surintendant. Their function was the supervision and verification of the actions of other financial officials, and they had seats on that part of the royal council known as the *conseil des finances*. Like the provincial intendants, they held their office on simple commission, presumably revocable upon the wish of the sovereign, but actually they had purchased their commissions from the crown. Until March 1649 there were only four intendants of finance: Séraphin de Mauroy, Barthélemy Hervart, Denis Marin, and Pierre Gargan. The government created four more that month: Jacques Bordier, Jacques Le Tillier, Guillaume Bordeaux, and Etienne Foullé. This addition was partly a financial arrangement, for each candidate paid a much-needed sum of two hundred fifty thousand *livres* to the treasury. But the government needed more fiscal agents because it planned on sending them on missions to replace the abolished provincial intendants.[14]

The career of one of these new intendants of finance offers an interesting example of the government's attempt. Etienne Foullé, seigneur de Prunevaux and son of a master of requests, served in the Parlement of Paris and as a president in the *cour des aides* in Guyenne before becoming a master of requests in his own right. Thereafter he acted as intendant in the generality of Limoges until the abolition of

13. Edmond Esmonin, "Un Episode du rétablissement des intendants: La Mission de Morant en Guyenne (1650)," originally in *Revue d'histoire moderne*, 1954, reprinted in *Etudes sur la France*, pp. 53–70. See also Charmeil, *Trésoriers de France*, pp. 380–84, on the use of masters of requests during the Fronde.

14. Julian Dent, "An Aspect of the Crisis of the Seventeenth Century: The Collapse of the Financial Administration of the French Monarchy (1653–61)," *Economic History Review*, 2d ser., 20 (1967): 241–56, is an excellent attempt in English to define the workings of the French department of finance; see p. 245 for the intendants of finance. Charmeil, *Trésoriers de France*, pp. 256–60. See also the anonymous *Estat et gouvernement de France: Comme il est depuis la majorité du roy Louis XIV, à present régnant*, 7th ed. of 1648 version (Amsterdam, 1653), pp. 84–85.

that intendancy in 1648. He then purchased the office of intendant of finance, and the crown returned him to Limoges in that capacity. There he claimed the right to preside over the local bureau of finances, but the *trésoriers* replied that as intendant of finance he had his place in the royal council and not in the generality, and if he claimed to be acting as a master of requests, he had no right to interfere in the affairs of the bureau.

When the *trésoriers* refused to have anything to do with him, Foullé quartered troops on their property, and his opponents claimed he even stooped to printing a false ordinance claiming to be from the bureau and added the names of its members. In addition to the "forgery," he intervened in the apportionment of the *taille* and obtained an order from the royal council suspending five treasurers in the bureau. Unsatisfied, he contemplated imprisoning the recalcitrant officials, but contented himself with announcing that with the cooperation of the amenable treasurers, he would levy the taxes. Next the Parlement of Bordeaux denounced his strong-arm tactics in collecting back taxes, but he continued to employ a band of archers, confiscating property without formal trial. Neither Foullé nor his assistants bothered to appear before the parlement to answer these charges.[15]

Foullé's example is not exceptional. The government used the intendants of finance on any mission needing capable royal agents. In 1650 Hervart arrived in an army camp in Germany to work with Philibert Baussan, intendant of Alsace, in negotiating with its officers for the payment of wages. That same year, Le Tillier journeyed to Dauphiné to hasten the payment of the *taille;* he had the authority to send armed men into the places that refused to pay. The year 1651 found Hervart in Dauphiné supervising the winter quarters of the soldiers lodged in that region. Near the end of the Fronde in 1653, Bordeaux traveled to Picardy to see to the payment of the troops in winter quarters and to ensure that they were ready for the approaching campaign. Marin did the same in Tours and Orléans, Gargan in Champagne, and Foullé in the Bourbonnais.[16]

The government also used army intendants to fill the vacancy left by the provincial intendants. This did not mean that the government

15. Charmeil, *Trésoriers de France*, pp. 259n136, 385–87.
16. Instructions given to M. Eruard (Hervart), 20 January 1650, A.G. A¹ 120, pp. 80r–87v; the king to Lesdiguières, mentioning Le Tillier, 7 March 1650, A.G. A¹ 120, p. 163rv; the king to Heruard (Hervart), 14 February 1651, A.G. A¹ 126, p. 54rv. Concerning Marin, Bordeaux, and Foullé, see the king's orders, 1–2 April 1653, A.G. A¹ 139, pp. 130r–33r. Commission for Gargan ("Sr de Bordeaux" is written in the margin) to levy forage in the generalities of Amiens, Soissons, and Châlons, 30 August 1653, A.G. A¹ 138, fol. 268.

always recalled army intendants serving on the frontier; sometimes it used the unemployed provincial intendants as army intendants, giving them a new commission to mask their real purpose. François Ville-montée, for example, was intendant of Poitou in 1648. Following the decision to suppress the intendants, he was recalled to Paris, and the next year he found himself serving as army intendant with the troops stationed in the Parisian suburb of Saint-Denis under the command of Du Plessis Praslain. Claude Bazin de Bezons was intendant at Soissons in 1648, but that December he replaced Pierre Goury as intendant of the army of Catalonia, a position he was not destined to hold long, for in 1650 the king sent him to the army inside Berry, where he served for two years. One final example might be cited: Jacques de Chaulnes was intendant of Limousin in 1648 when he received his letter of recall. In April of the following year, he became intendant of justice, police, and finance in the army of Flanders under the command of the comte d'Harcourt.[17]

Sometimes these army intendants reinforced their authority by adding other titles. When Louis Laisné, seigneur de la Marguerie, and Nicolas Fouquet were sent as joint intendants of justice, police, and finance to the army in Normandy, they combined this office with that of a master of requests riding on circuit. Denis Marin is another example; when he received a commission as intendant of finance in the army of Guyenne, he was already an intendant of finance. In 1650 Jacques Paget was intendant of the army of Champagne and at the same time a master of requests on *chevauchée*, which allowed him to watch civilian officials.[18]

17. Secretarial footnote to the commission of Choisy, 15 January 1649, noting that a similar one had been sent to Villemontée for the army in the quarter of Saint-Denis outside Paris, A.G. A¹ 114, pp. 28v–29r; commission of Bezons as intendant of the army of Catalonia, 16 December 1648, A.G. A¹ 108, pp. 232r–34v; almost illegible commission of Bezons as intendant in the army of Berry, 25 October 1651, A.G. A¹ 125, fol. 196 (Bezons had also served as army intendant in Berry the previous year; see his letter to Le Tellier, from Bourges, 24 September 1650, A.G. A¹ 118, fol. 202, and Loménie de Brienne to the marquis de La Vieuville, 23 October 1651, A.G. A¹ 127, pp. 86r–88r: "l'on mande M Bezon de venir icy faire lintendance po. les troupes come il a fait tres dignement l'année passée et que cette resolution a été prise sur le peu de satisfaction qu'on eu des tresoriers de France de cette gñalitez."). Secretary's notation that a letter had been sent to Chaulnes, 15 July 1648, recalling him from his intendancy in Limousin, A.G. A¹ 108, p. 12v; rough draft of a commission for Chaulnes as intendant in the army of Flanders, dated "dernier avril 1649," A.G. A¹ 113, fol. 201.

18. Charmeil, *Tresoriers de France*, p. 389; commission of Fouquet as intendant of the army of Normandy with a secretarial notation concerning Laisné, A.G. A¹ 120, pp. 104r–5v; Marin's commission as intendant of finance in the army of Guyenne, 25 September 1651, A.G. A¹ 126, pp. 471r–73r. Concerning Paget, see his instructions as army intendant in Champagne, A.G. A¹ 120, pp. 156r–59v, and the reference to his *chevauchée* in Esmonin, *Études sur la France*, p. 55.

The office of provincial intendant had become so indispensable that the crown was not the only one to use army intendants to enforce its authority. The Frondeurs, too, named army intendants to collect taxes in the provinces and to supply the rebel forces. In Guyenne alone there were five who functioned as rebel army intendants–cum–provincial intendants: Duduc from Mas d'Agenais; Florimond de Reymond, who had close relations with Conty; Mosnier, former *lieutenant criminel* of the *élection* of Guyenne and father-in-law of the Ormist *enragé* Pierre de Villars; D'Espagnet, remembered for his devotion to Condé and his incorruptible virtue; and finally the most infamous of all, Jacques de Guyonnet, who plundered the countryside to provision Condé's army.[19]

The army intendant's most important mission was to collect the *taille*, and surely the pretensions of the rival intendants did nothing except confuse the bewildered peasants, who were already hard-pressed to survive let alone pay two different tax levies. Nevertheless, each intendant, royal and parlementarian, resorted to decrees forbidding the payment of taxes to the enemy. One such printed decree still exists. Issued by Pierre de Pontac, intendant of the comte d'Harcourt's army in Guyenne, it denounced the disorder in the province caused by the decrees of the rebellious Parlement of Bordeaux and its "legion of intendants." This circular particularly mentioned Mosnier, who had tried to seduce the populace "on his own authority" by persuading them that if they paid three-fourths of their *tailles* to him, he would release them from the obligation to pay the remaining quarter. After reaffirming the order for the royal tax collectors to use "all the diligence necessary" in the recovery of money due from last year's unpaid taxes, the document ended with this warning: "We do break and have broken all judgments, agreements, and ordinances rendered on the *tailles* and moneys of His Majesty by the said commissioners and pretended intendants," and it expressly forbade any of the king's subjects to comply with their demands, "on pain of punishment." [20]

---

19. L. Couyba, *Etudes sur la Fronde en Agenais* (Villeneuve-sur-Lot, 1901–3), 3:47–50. Not all the men in Guyenne were, strictly speaking, intendants of the army. For example, in the Archives Départementales de Lot-et-Garonne at Agen, there are various documents in Series EE.21, Ville d'Agen, unpaged, that mention "Mr de Guyonnet coner et intendant," "Jean duduc coner du Roy en la cour de parlemt de Bourdx commre par Elle depute et commre genal des armees du Roy soubs l'authorité de Son Altesse," and "le sieur de Moynier coner du Roy en sa cour de Bourdeaux commre par Elle deputé Intendant de larmee de Sa Mate sous l'autorité de Son Altesse." One would need to examine their commissions, if any are still preserved, before one could make a more accurate designation.

20. Printed circular of Pontac, dated "d'Agen le 25 mai 1652," A.G. A¹ 133, fol. 358.

To give support to such circulars, each side resorted to writs from friendly parlements to reinforce the legality of its cause. The Parlement of Toulouse, for example, supported the crown by issuing an *arrêt*, or warrant, on 25 November 1651 declaring the rebel intendant De Guyonnet guilty of lese majesty and forbidding all loyal subjects to recognize his authority or pay taxes to him. Yet royal power remained impotent against these rebel intendants in areas like Guyenne, long a stronghold of the Condé family. Unless the royal armies were at their gates, towns preferred to treat with both sides to avoid the consequences of offending either; the peasants, as usual, continued to suffer.[21]

Disloyalty to the crown was as important a problem as tax collection. Disaffection appeared among both civil and military officials— the rebellious parlements as well as the princely Fronde—and there seemed to be little short of force that the government could do to establish order. Again it was the army intendant who played a role in solving this problem. For example, Pontac, intendant of the royal army in Guyenne, wrote frequent letters to the crown recommending methods for reducing the insurgents: Bordeaux would submit to royal authority only after military or economic pressure; less important towns should be infiltrated by persons loyal to the king. Pontac described the municipal officers of Agen as "creatures of Monsieur the Prince" (Condé) and urged the rigging of a town election in favor of the government. The election of the city council of Agen always took place on 31 December, and if the court approved the nominees, Harcourt could fix the election. Pontac remarked: "I do not see any other guarantee than this of preventing the consuls, who are presently in charge and who have merited the hangman's noose a hundred times, from naming and substituting in their places persons as dangerous as themselves."[22]

Although rebellion in the provinces occupied an army intendant's

21. Jacques de Saulx, comte de Tavannes, *Mémoires de Jacques de Saulx comte de Tavannes suivis de l'histoire de la guerre de Guyenne par Balthazar*, ed. Célestin Moreau (Paris, 1853), p. 293*n*3. For a description of the courtesy visit that the consuls of Agen made to the rebel prince of Conty and intendants D'Espargnet, Mosnier, and De Guyonnet, see Couyba, *Etudes sur la Fronde en Agenais*, 2:100.

22. Concerning Pontac's plan to place an economic boycott by naval blockade across the mouth of the Garonne River to prevent the exit of wine and the import of grain, see his letter to Mazarin, 19 September 1652, from Agen, in Gabriel-Jules de Cosnac, ed., *Souvenirs du règne de Louis XIV* (Paris, 1876), 5:67–71. Pontac to Le Tellier, 19 October and 27 November 1652, A.G. A¹ 137*bis*, fol. 457, and A¹ 134, fol. 396. Pontac also worried about the adherents of Condé who fought with the prince during the campaign season, then quietly returned home where they continued to foment rebellion; he suggested that they be forced to take oaths of fidelity before they could enter the towns held by the king; see his letter to Le Tellier, 4 November 1652, A.G. A¹ 134, fol. 354.

attention, disaffection in the army posed a more pressing problem. Unless the crown could rely upon its troops, no effort could be made to pacify the countryside or continue the war against the foreign enemy. The war department's dilemma lay not so much with Frondeur princes inside France as with the mass of troops, unpaid and lacking forage and supplies. The soldiers turned to looting, and after indulging in this indiscipline, were more reluctant than ever to obey. Their impoverished conditions forced the officers to side with their troops, and even some intendants sympathized with the soldiers' complaints. Charles Machaut, provincial intendant in Burgundy—one of the reserved intendancies—protested the government's orders for winter quarters in 1649, particularly the decision to pay only two *môntres* in wages instead of the four or more needed if the garrisons were to live in peace with the populace. Although he promised to punish army disorders to the best of his ability, Machaut warned that soldiers had been established by God for the defense of kingdoms and the rights of sovereigns. Even Jesus, he piously reminded the crown, had allowed his followers to pluck grain as they passed through fields in order to satisfy their hunger.[23]

Other intendants faced the same problem of unpaid and undersupplied troops and their resulting disaffection. Ennemond Servien, the intendant-ambassador in Italy, reported in 1649 that the outcry of the Amboise regiment over the shortage of provisions was so great that the regiment had been permitted to leave Turin. Its officers had boldly declared that unless this were done, they would take up arms against the garrison if the town was besieged. In Catalonia in 1652, army intendant Aligre reported that the troops lacked everything: money, cannon, and supplies. The troops were at the point of open rebellion, muttering that they would serve whoever did the most for them. And the general did nothing to silence their complaints. In northeastern France, Philibert Baussan, who had taken leave from his frontier intendancy of Alsace to serve as army intendant, reported that the cavalry officers had left their assigned quarters complaining that they were accustomed to better treatment. The intendant added: "It seems that they have some *mauvais* design, especially because of rumors of the disbanding of regiments that are current here. I hope that all these bad humors dissipate."[24]

23. Machaut to Le Tellier, from Dijon, 17 December 1649, A.G. A¹ 116, fol. 482. In the letter, Machaut described the troops as naked (*tous nuds*) and moralized: "Il couste moins au peuple de donner largement avec ordre et police aux gens de guerre leur besoins que les obliger par necessité a les prendre avec violence et rapine."

24. Servien to Le Tellier, 11 December 1649, A.G. A¹ 116, fol. 470; Aligre to Le Tellier, 15 March 1652, describing the troops in these terms: "avec leur in-

Rumors of impending revolts often proved well founded. For example, German mercenary cavalry serving under the command of Rosen mutinied in 1650. Le Tellier confessed to Antoine Bordeaux, intendant of the army in Flanders, that the insurrection was *très fascheuse*, but that Rosen promised henceforth to maintain his troops in good order. The government was willing to cooperate with him because it needed his services, but the secretary warned the intendant that it was important to continue the search and punishment of those involved in the uprising, and that an example should be made of the major accused of leading the conspiracy.[25]

The government suspected many of the aristocratic army officers in the period preceding and during what is known as the Princes' Fronde. After the surrender of a frontier garrison without a struggle in 1649, the government dispatched an army intendant to investigate. Harcourt had decided to besiege Cambrai and camped his army before its walls. Unfortunately for the French, the Austrian archduke had camped at Douai, where he could easily relieve Cambrai. Profiting from a thick fog, he attacked the French forces, throwing two thousand men into the city and lifting the siege. It was a humiliation for the French government. Mazarin, beside himself with rage, mistrusted Harcourt, especially when he learned that an intercepted enemy letter should have given the general ample warning. Even stranger was the fact that the Spanish had entered by a gate left open for forage, supposedly guarded by fifteen men who mysteriously had left their post at the time of the attack. The crown immediately sent a letter to Jacques de Chaulnes, intendant of the army, requesting that he inform the secretary of the place where the enemy attacked, the number of men and their commanders, whether any resistance had been made, and, in brief, "everything which passed on this occasion." Le Tellier emphasized that the information should be sent closed and sealed, and that he was relying on Chaulnes's diligence and fidelity.[26]

The loyalty of a lieutenant general, Marsin, long a friend of Condé, also did not go unquestioned when that prince waged open warfare with the government. The king sent a letter to young Simon Arnauld

solence faisans de grigues et jettant des billets," A.G. A¹ 137, fol. 57; Baussan to Le Tellier, 7 December 1649, A.G. A¹ 116, fol. 462. Parker, *Army of Flanders*, p. 219, remarks that "reforming" or disbanding companies was extremely unpopular in armies, particularly among the officers, for their positions were suppressed. Often such a "reformation" was used to punish the troops and to redistribute them among more loyal companies.

25. Le Tellier to Bordeaux, 21 October 1650, A.G. A¹ 120, pp. 504v–6r.

26. Pierre-Adolphe Chéruel, *Histoire de France pendant la minorité de Louis XIV* (Paris, 1879), 3:265–67; the king to Chaulnes, 6 July 1649, A.G. A¹ 115, p. 2rv.

d'Andilly, the future marquis de Pomponne and intendant of the army of Catalonia in 1651, warning: "you are to watch his conduct and have the governor of Catalonia and some of my most faithful servants whom you feel you can trust with a matter of this importance to be on guard, so in case that you see that he has sinister intentions against my service, you can prevent him by any means from putting them into effect." The crown's suspicions were well founded, because in September of 1651 Marsin deserted the army to join Condé in Auvergne, taking a large body of troops and part of the treasury with him. Although the government sent instructions to Marca, the visitor general, to arrest Marsin and imprison him in the citadel of Perpignan, the order arrived too late to be executed.[27]

Army intendants rarely made important arrests, probably because they had too little authority or prestige to jail a high aristocrat. During the Fronde, the secretary of war generally addressed such orders to other army officers. In 1649 General Erlac received instructions to arrest a sieur Tracy, and the count de Paluau the same for the arrest of Marshal Rantzau. Only in Italy did Servien, the intendant-ambassador, have power to arrest a lieutenant general. The crown used the army intendants to imprison lesser officials, for Pomponne arrested an assistant to the treasurer of war who was caught with Marsin's secretary trying to embezzle money destined for the payment of troops. But Pomponne did not try the criminal. He was sent to Breteuil, intendant of Languedoc, for trial before the Parlement of Toulouse.[28]

Servien seems to have been an exceptional case. Documents in the war archives reveal that the intendant-ambassador played a significant role in two important cases, one civil in nature, the other treasonable. In the first, a Captain La Londe, commander of the fort of Perouze, participated in an illicit salt trade to defraud the tax farm in Dauphiné.

---

27. The king to Simon Arnauld d'Andilly, 11 September 1651, A.G. A¹ 126, pp. 426v–27v. The crown had good reason to suspect Marsin; he had previously served in Catalonia in 1649, but when Condé joined the Fronde, Marsin sided with him. After Condé's release from prison in 1651, his followers were pardoned, among them Marsin, who returned to his command in Catalonia; see the king to Marca, bishop of Coserans and visitor general, announcing the return of Marsin to Catalonia, 12 May 1651, A.G. A¹ 126, pp. 174r–77v. Marsin remained in Catalonia until 28 September 1651, when he once again deserted to join Condé, taking four regiments with him. See Henri d'Orléans, duc d'Aumale, *Histoire des princes de Condé pendant les XVIe et XVIIe siècles* (Paris, 1892), 6:113–14.

28. The king to Erlac, to arrest Tracy, 16 February 1649, A.G. A¹ 114, pp. 107v–8r; to Paluau, to arrest Rantzau, 19 February 1649, A.G. A¹ 114, pp. 115v–19v; to Breteuil, 22 November 1651, A.G. A¹ 127, pp. 211v–12r. The Tracy mentioned is almost certainly Alexandre de Prouville, sieur de Tracy, who served in Germany from 1641 to 1649. He commanded a regiment and was also a commissary general in the army. Later, 1663–67, he served as lieutenant general of Canada.

Although it was the governor of Pignerol who arrested La Londe, the crown sent Servien orders to conduct the trial by virtue of his commission as intendant of justice, police, and finance. Unfortunately the outcome is unknown; only a letter to Servien in February 1650, instructing him to continue the trial, exists.[29]

The second case was more serious. The government suspected the loyalty of Saint-Aunetz, lieutenant general in the army of Italy, and ordered Servien to direct the capture and trial. For this purpose, Le Tellier provided him with letters to the sieur de Monty and the marquis de Ville, both brigadiers, to seize the lieutenant general. Servien, however, held the key position in the arrest since he had an alternate set of instructions, "as there could be some inconvenience in concerting the thing with both of them, which cannot be foreseen from here [Paris]." If Servien decided that it was impossible to arrest Saint-Aunetz in the army, he was to arrange the matter in Turin, Suze, or Pignerol with or without notifying the brigadiers—since he had a set of orders for the officers ignoring the arrest, merely instructing them to take command of the army and to obey orders. Although Le Tellier warned Servien to act with caution, he encouraged him to confer with Simon Arnauld d'Andilly, who had formerly served as intendant of the naval expedition to Italy in 1648 and of the fortress of Casal, "whom His Majesty values for his ability and adroitness—*son intelligence et adresse*." Although the procedure followed is unknown, Saint-Aunetz was arrested and imprisoned in the fortress of Pignerol until his release in April 1650, when, like so many other Frondeurs, he was pardoned after the first Condéan rebellion.[30]

Although army intendants usually did not play an important role in arrests, one document indicates that they could be actively involved in trials—if the crown so desired. For example, the instructions given to Jacques Paget when he became intendant of La Ferté Senneterre's army in Champagne at the beginning of 1650 ordered him to investigate charges against various prisoners involved in Turenne's participa-

29. The king to Servien, 6 November 1649 and 10 February 1650, A.G. A¹ 115, pp. 135v–36v, and A¹ 120, pp. 138v–39r.

30. Memoir for Servien concerning the arrest of Saint-Aunetz, 22 November 1649, A.G. A¹ 115, pp. 165v–69r; the king to M. de Malissy, to release Saint-Aunetz from the citadel of Pignerol, 20 April 1650, A.G. A¹ 120, p. 221rv. Servien had instructions to try Saint-Aunetz "en vertu du pouvoir porté par la commission que vous aves eue cy devant dintendant de la justice en mon armée d'Italie et dela les montes, procedant contre luy pour raison des desseins contre mon estat et mon service. . . . jusques a jugement definitif exclusivement et que dans toutes votre procedure vous gardies les formes en pareil cas requises et accoustumees en conformité de votre commission"; the king to Servien, 4 January 1650, A.G. A¹ 120, pp. 5r–6v. In this and other passages, I have translated the office of *maréchal de camp* as brigadier, following *New Cassell's French Dictionary*, s.v. "maréchal."

tion in the Fronde. The prisoners included those in the garrisons of Stenay, Domvilliers, and Clermont; the companies of *gendarmes* and light cavalry of Condé, Enghien, and Conty; the cavalry regiments of Duras; and the infantry regiments of Turenne. Paget's instructions called for him to visit the places where the prisoners were guarded, interrogate them, and hear the testimony of all available witnesses. Then he was to conduct their trials, together with judges from the nearest *sièges présidiaux*, judging them "according to the form and rigor of the law." [31]

Unlike the alienated Frondeur officers, the army intendants remained loyal to the government, both in their temporary tasks as substitute provincial intendants and in their more traditional activities of supplying armies, making contracts with munitioneers, paying troops, and even borrowing on their own credit to aid the army. Not one army intendant was arrested during the Fronde. The explanation for this fidelity lies in two ideals: devotion to his master and an esprit de corps. A remark of Jacques Hector de Marle, seigneur de Beaubourg and intendant in Lorraine, although not strictly speaking an army intendant, illustrated the devotion ideal. He had charged the *sénéchaussée* of Remiremont with the expense of one company for winter quarters, despite the fact that this territory was under the protection of the duchess of Orléans. She had protested vigorously to the intendant, but with no effect, for as Beaubourg wrote Le Tellier, "when you send me contrary orders, I will do it, for I owe you obedience." The intendants also had a common sense of purpose—perhaps not as great as the corporative cohesion of the officeholders, but with a higher ideal than their own self-interest. The words *le service* and *pour le service du Roy* ring repeatedly throughout the intendants' letters. This was not a new development during the Fronde—the intendants had used such phrases in the 1640s—yet undoubtedly the ideal knitted them closer to the crown during these troubled times.[32]

This esprit de corps bound the army intendants together when they

---

31. Paget's instructions, 27 February 1650, A.G. A¹ 120, pp. 156r–59v.

32. Beaubourg to Le Tellier, 24 December 1649, A.G. A¹ 116 fol. 495. Instances of the term "le service" are numerous; see, for example, the letter of Pierre Goury, 26 June 1648, A.G. A¹ 106, fol. 80: "me promet bien que celles [his illness] ne retardera en rien le service"; and the remarks of Servien to Le Tellier, 23 November 1649, A.G. A¹ 117, fol. 373: "j'aprenhenderois quelque alteraõn au service du Roy si ie faisois quelque entreprise de mon chef." Certain creatures were so subservient to Le Tellier that they consulted their master before embarking on what might be regarded as minor tasks. Thus Goury asked Le Tellier whether it would be appropriate for him to write a congratulatory letter to La Meilleraye on his nomination as superintendent of finance, promising: "j'attendray den faire ce que vo, me prescriverez," 3 August 1648, A.G. A¹ 106, fol. 123.

felt their rights threatened or when they were encroaching upon the prerogatives of others—the distinction between the two being very thin in the seventeenth century. For example, the year 1651 saw a conflict between the intendants of the army and a rival group of officials, the superintendents–commissary generals of supplies. The struggle broke into the open with a disagreement between Herbert, the superintendent–commissary general, and Bordeaux, intendant of the army in Flanders. The government tried to work out a compromise by its ordinance of 12 May 1651. This document noted that the army intendants, under the pretext that royal regulations required them to have knowledge of all the expenses of the army relating to supplies ordered by the generals, were daily countersigning (*visent*) by themselves the orders of the generals. The crown now decided that the army intendants and the superintendents–commissary generals should jointly cosign all such documents in the future, although in the absence of one the other's action would be sufficient.[33]

33. Ordinance of 12 May 1651, A.G. A¹ 126, pp. 179v–81r. Although the office of *surintendant et commissaire général des vivres* dated from the days of Louis XIII—it had been created in 1627, then suppressed twice, once in 1631 and again in 1635—there are few documents relating to it in the French war archives. The fragmentary correspondence of several of these officials, Tracy in the armies in Germany, Le Clerc in Catalonia, and Herbert in Flanders, does not indicate that they were a regular institution, although the office had been re-created by an edict of August 1643, and supplements of 30 April and 15 December 1644. The position was frequently vacant, and in this vacuum the army intendants took over their function. Furthermore, the duties of the superintendents–commissary generals were deliberately vague and broad. The edicts of the early 1640s, for example, stated that they were to have "the intendance and entire direction of supplies in the armies. . . . regardless of all commissions dispatched to the intendants of justice and finance in the armies and other commissions, articles, and contracts to the contrary awarded to the munitioneers." See André, *Michel Le Tellier et l'armée*, pp. 446–55.

Little has been written on the office of *surintendant et commissaire général*, although André thought it was part of a major effort of Le Tellier to direct the supply of the armies, remarking that Le Tellier "accorde une importance chaque jour plus grande aux surintendants commissaires généraux des vivres" (*Michel Le Tellier et l'armée*, p. 446). André defined the office as exercising a control over everything regarding the soldiers' subsistence. The commissary generals watched the munitioneers to see that they fulfilled their contracts, reviewed their equipages, and inspected their accounts. They supervised the bakers, wagoners, and guards, but their most important mission was the amassing of supplies and their transportation to the army—the *voiture des vivres*. However, I remain unconvinced that they were such a permanent and important institution as André implies. In the documents I examined in the war archives, dating from the period between 1630 and 1670, I found only passing mention of these officials; mostly the problem of supply involved the army intendant, the *commissaires des guerres*, or the provincial intendant.

François Nodot, a former *commissaire des armées*, wrote *Le Munitionaire des armées de France, qui enseigne à fournir les vivres aux troupes avec toute l'oeconomie possible* (Paris, Brussels, and Lyon, [1697]), which mentions only a *général*

The army intendants seem to have ignored this decree, for three months later, on 22 August 1651, the government issued another ordinance mentioning the complaints on each side. Bordeaux affirmed that the intendants before him had signed such papers by themselves and had had the direction of supplies. Thus he could not cosign without prejudicing his position. Herbert replied that according to the edicts reestablishing his office, he was to have the direction of supplies and it was a diminution of the honor and function of his charge to have the intendant cosign his papers. Furthermore, since Bordeaux made difficulties about cosigning, the king should permit the super-intendant–commissary general's signature alone. The crown decided to reaffirm the ordinance of 12 May, with the provision that if Bordeaux refused to sign jointly, Herbert's signature would be sufficient. This did not seem to solve the problem, because as late as 1653 the war department addressed similar orders to Brachet, intendant in the army of La Ferté Senneterre in Italy. Further research needs to be done on this problem, but the intendants continued to have a significant role in the supply of armies.[34]

The period of the Fronde was also a time of delicate relations between the army intendants and the generals, although the secretary of war continued to warn the intendants, as usual, to submit to the generals' wishes so there would be no reason for disagreement. But this period of parlementary and princely resistance to royal authority made the aristocratic army officers more sensitive than ever to a pretended slight to their prerogatives. For example, during Le Tellier's brief exile to his estates in 1651, his substitute in the war department, Loménie de Brienne, found it necessary to reprimand Ennemond Servien, intendant

des vivres, but this agent was appointed by the munitioneer and was "un ancien commis homme experimenté," that is, an experienced subordinate able to take charge "si le munitionaire ne peut aller lui-même à l'armée (p. 137). The last chapter of the book, entitled "De la manière ancienne de faire les vivres en France," speaks of the office of *sur-intendants et commissaire généraux des vivres* in Louis XIII's time, but remarks: "Tous ces officiers n'ont plus maintenant d'exercice, depuis qu'on a déchargé les peuples de la contribution des vivres, et qu'on les a donnez a fournir par entreprise à des traitans, qui les font exercer par commission" (p. 608). Charles-Nicolas Dublanchy, *Une Intendance d'armée au XVIIIe siècle: Etude sur les services administratifs à l'armée de Soubise pendant la guerre de sept ans* (Paris, [1905]), pp. 42–47, is in agreement, remarking that the supply of armies in the eighteenth century was by private enterprise. The consortium of munitioneers had their office in Paris and appointed their own man in the field, who had the title of *directeur général des vivres* but was commonly known in the army as the *munitionnaire.* There does not seem to have been a superintendant-commissary general.

34. Ordinance regulating the differences between Herbert and Bordeaux, 22 August 1651, A.G. A¹ 126, pp. 391r–93r. The king to Brachet, to conform to this ordinance, 27 July 1653, A.G. A¹ 140, pp. 92v–93r.

in the army of Italy: "Prince Thomas complains that you alone order the distribution of the bread rations in the army. Because you know that this is a thing contrary to what is practiced, which is that the generals or lieutenant generals order and the intendants countersign their regulations and they [the intendants] order only where there are brigadiers; I think that you should conform to this [practice]." [35]

Beaubourg, intendant on the Lorraine frontier, reported similar conflict with La Ferté Senneterre, who wanted to receive eighty-two rations for his staff—his *état-major*—as had been ordered for the German mercenaries, instead of the thirty-two customary in the French army. Beaubourg remarked that he could not prevent the general from doing what he wanted, but he would refuse to approve it until he received orders to the contrary. "Often these little difficulties," Beaubourg concluded, "rise between us about things that concern his interests." The intendant, however, promised to abide by Le Tellier's instructions.[36]

Because of the nature of an intendant's duties, generals felt it important to have a sympathetic man in this office. Jean-Edouard Molé, sieur de Champlastreux, is a good example. A close friend of Condé, he served as his army intendant on numerous campaigns: in 1644 in the army on the frontier of Champagne, again in 1646 in Condé's army on the northern frontier, and with Condé in Catalonia in 1647. Obviously, part of the reason for Champlastreux's continual service near Condé was that the prince found him to his liking, and one did not ignore the wishes of such a powerful man. Of course it did not hurt Champlastreux to be a friend of Mazarin. It was often important to have a foot in both camps, particularly during the Fronde. Similar mention might be made of Jean de Choisy, a long-time provincial intendant in Champagne and friend of the crown. He had served as army intendant on numerous occasions and obtained the office of chancellor in the duke of Orléans's household, where the duke sometimes used him for sensitive personal business. It is not surprising that he obtained a commission as intendant of the army commanded by Orléans in 1649.[37]

During the Fronde, when the government sought to retain the support of important princes, it was only natural for the generals to

35. Brienne to Servien, 11 August 1651, A.G. A¹ 126, pp. 377v–78v.
36. Beaubourg to Le Tellier, 21 December 1649, A.G. A¹ 116, fols. 486–87.
37. Regarding Champlastreux, see his commissions: A.G. A¹ 86, fol. 130; A¹ 96, fol. 143; A¹ 103, pp. 124r–26r. Pierre-Adolphe Chéruel, ed., *Lettres du cardinal Mazarin pendant son ministère*, Collection de documents inédits sur l'histoire de France (Paris, 1872–1906), 4:26*n*5, remarks that Champlastreux "était particulièrement attaché à Condé." Commission of Choisy, 15 January 1649, A.G. A¹ 114, pp. 26v–29r; for a brief biography, see *Dictionnaire de biographie française*, s.v. "Choisy, Jean Iᵉʳ et II de."

play a more active role in the selection of army intendants. Sometimes they took the nomination out of the crown's hands. In 1652, when Denis Marin, the intendant of finance, accompanied the comte d'Harcourt's army, necessity forced him to remain in Saintonge to prepare the siege of Saintes while Harcourt moved southward. On 28 February the general boldly announced to Le Tellier: "We have judged it necessary in his [Marin's] absence to give the handling of finance in the army to Monsieur de Pontac, first president of the *cour des aides* in Guyenne, whose probity and credit has been proven to us." Harcourt had nominated his own intendant and now asked the government to confirm his choice. Although no general had the right to appoint an intendant, the war department agreed to Harcourt's wishes.

Indeed, it had little choice. The regular intendant was absent, but according to Harcourt's letter, Marin had encouraged the choice. A local man on the scene might prove valuable, although his fidelity was an unknown quantity. The government sought to limit Pontac's power by delaying his commission for almost two months and by restricting his authority to that of intendant of finance. Harcourt immediately wrote back to the war department on 30 April 1652: "The commission you have taken the trouble to send me for President Pontac only gives him the direction of finances. Without doubt you will judge it appropriate [*fort apropos*] to send him a specific one as intendant of justice and police in this army, in which he has up until now performed the duties with a probity known to you. In such a case, I think it unnecessary to remind you that it [the commission] must not be limited to Monsieur Marin's absence, as was that of finance." [38]

The situation in Guyenne took a dangerous turn when Harcourt, smarting under the shame of a military defeat at the siege of Villeneuve, fled to his estates in Alsace on 16 August 1652. The flight of his patron rendered Pontac's position insecure, but he tried to justify himself by declaring that he had no foreknowledge of the general's flight. The war department had no proof of Pontac's complicity, and it was under some obligation because Pontac had advanced considerable sums of his own money to supply the army. Furthermore, Pontac

38. Harcourt to Le Tellier, 28 February and 30 April 1652, A.G. A¹ 133, fols. 96, 253. There are two variant copies of commissions preserved in the war archives. The first, dated 12 April 1652, named Pontac as intendant of justice, police, and finance, but the copy must be in error. The other commission, dated 18 May 1652, corresponds to the second commission requested by Harcourt, for it named Pontac intendant of justice and police; A.G. A¹ 135, pp. 183r–84r, and A¹ 132, fol. 303. That originally only a commission as intendant of finance had been dispatched is confirmed by a clerk's notation on the back of Harcourt's letter of 28 February: "a commis M. Pontac intendant de fi n [*sic*] armee en la place de M. Marin."

urged, "the knowledge that I have of the affairs in this province and of the tax collections that are being made, give me an advantage over any others." Le Tellier allowed Pontac to keep his intendancy until November 1652, but he stripped the intendant of most of his authority and surrounded him with Marin and Tracy, a commissary general. Pontac bitterly remarked: "having consented to please you that the sieur de Tracy have the entire handling of finances, I find that I am the only one who has not been paid his salary." Thus Pontac was gradually eased out of his intendancy.[39]

Harcourt was not the only example of a general's role in the selection of an intendant. After Pontac's departure, the war department intended to replace him with a sieur de Thyersault, a master of requests, but a letter from Harcourt's replacement, the duc de Candale, remarked that Thyersault was not to Tracy's or his liking —"n'est nullement à son goust ni au mien"—and recommended the appointment of Ribeyre, a *maître des comptes*. In due course Ribeyre suceeded Pontac as intendant of the army in Guyenne.[40]

Finally there is the curious example of the duc d'Elbeuf, who selected an intendant, then changed his mind and attempted to remove his choice. Elbeuf was governor of Picardy and also one of the generals stationed along the frontier. He had requested Geoffrey Luillier d'Orgeval as intendant; but soon afterward, Le Tellier received a letter from the general that bluntly stated: "I am forced to confess to you that I made a very bad choice [*meschant chois*] of the person of M. d'Orgeval when I proposed him to the queen as intendant of Picardy." Although Elbeuf admitted that he had received all the marks of submission that might be desired, the intendant was a man "without a brain [*sans cervelle*] and without order who spends three weeks on the same affair without concluding anything." In brief, he

39. Pontac to Le Tellier, justifying his conduct after Harcourt's flight, 16 August 1652, A.G. A¹ 134, fol. 229. To prove his innocence, Pontac enclosed the general's note to him before his flight; this letter, Harcourt to Pontac, 16 August 1652, is preserved in A.G. A¹ 134, fol. 208. Concerning the orders for Pontac to submit to Tracy and Marin in matters of finance, see Pontac's remarks to Le Tellier, 19 November 1652, A.G. A¹ 137*bis*, fol. 495; there is similar information in Pontac to Le Tellier, 4 November 1652, A.G. A¹ 134, fol. 345, and in Pontac to Mazarin, 19 September 1652, in Cosnac, *Souvenirs du règne de Louis XIV*, 5:67–71. Marin spoke of Pontac's connection with Harcourt in this manner: "intendant et son grand confidant," letter to Le Tellier, 30 August 1652, A.G. A¹ 134, fol. 289. Cosnac, *Souvenirs du règne de Louis XIV*, 3:149–50, speaks of Pontac in this fashion: "mais la disgrâce qui le frappait était commune à tous ceux qui avaient servi sous le comte d'Harcourt, la défection du chef les avait rendus suspects."
40. The king to Pontac, recalling him as intendant, 8 November 1652, A.G. A¹ 136, pp. 393r–94v, mentions the intention of making Thyersault intendant, but see the duc de Candale to Le Tellier, 14 November 1652, A.G. A¹ 134, fol. 342. Ribeyre's commission is in A.G. A¹ 132, fol. 304, dated 25 November 1652.

was the despair of the army and the wrong should be remedied immediately.[41]

Elbeuf suggested that Luillier d'Orgeval be replaced by a man whom, the governor stressed, he did not know but about whom he had heard good things. The man was Monsieur Vauxtorte, and Elbeuf promised to receive him "with open arms." Despite Elbeuf's recommendation, the government temporized. Orgeval was sent on another mission, leaving Julien Pietre, treasurer general of France at the bureau in Amiens, to act as intendant during his absence. Although Elbeuf thanked Le Tellier for removing "the most malicious and dangerous of all the fools that I have ever known," the government was wary. Mazarin had suspicions about Elbeuf, for he remarked to his confidant, Ondedei, that the duke had written a slanderous letter against the cardinal to the marquis d'Hocquincourt.

Mazarin believed that Elbeuf and his son wanted to get rid of Orgeval as intendant not because they had anything against him, but because they wanted the post all along for Pietre, "their partisan, enemy of Orgeval." To thwart the scheme, Mazarin proposed Antoine Bordeaux as intendant. The latter had experience, but most important, Mazarin added, "he is entirely dependent upon me." Thus Orgeval was not disgraced by Elbeuf's efforts; the government conveniently shifted him to the provincial intendancy in Provence. And Elbeuf did not get his man; the crown had successfully asserted its prerogative of selecting the army intendants. The government had learned an important lesson from the Fronde. It needed royal agents on whom it could depend—loyal men such as Beaubourg who would say, "I owe you obedience." For this reason, it was vitally important that the crown keep the nomination and choice of intendants in its own hands.[42]

With the crown once more in control of France by 1653, it again turned its attention to the war with Spain. The lifting of the siege of Arras (1654) prompted the beginning of negotiations that after five lengthy years would culminate in the Treaty of the Pyrenees. The bulk of documents preserved in the French war archives reflects the trend to peacetime activity, both in number and variety. First, they decline in volume; the dispatches preserved for 1659 equal only about a third of those collected for a year earlier in the 1650s. Second, these documents are mostly concerned with internal pacification, instructions for winter quarters, distribution of troops within provinces, and demolition of rebel fortresses.

41. Elbeuf to Le Tellier, 18 April and 8 July 1652, A.G. A¹ 137, fol. 89, and A¹ 137*bis*, fol. 254.

42. Mazarin to Ondedei, 4 September 1652, in Italian, in Chéruel, *Lettres du cardinal Mazarin*, 5:200–204.

The activity of the army intendants was of minor significance compared to that of such provincial intendants as Charles Le Jay in Lorraine, François Bochart de Sarron Champigny in Lyonnais, François Villemontée in Soissons, and Claude Bazin de Bezons in Languedoc. Only a few documents exist for the army intendants: Talon in the army in Flanders, Jacques Brachet in Italy, and Louis Gombault in Catalonia. A brief survey of this fragmentary correspondence reveals little new about the office of army intendant. It was almost as if the war department as well as the country was recuperating from the turmoil of the Fronde. It would not be until the 1660s that the office of army intendant would assume new importance, and those years, oddly enough, would be years of relative peace disturbed only by several military expeditions of the young eaglet Louis XIV anxious to spread his wings.

VI

The Years of Preparation
1660-66

The end of hostilities with Spain meant a reduction in the armed forces. Even the last two army intendants had disappeared by 1659: Brachet died that year while on campaign in Italy, and the king named Talon as one of the special commissioners to settle the territorial boundaries in northern France in conformity with the Treaty of the Pyrenees. Since army intendants served only for the duration of a campaign, there was now no need for them. Yet the supervision of an army, even in peacetime, called for time and effort on the part of its administrators. The *commissaires des guerres* stepped into the place vacated by the army intendants, and their position assumed more and more importance during the decade of the 1660s. Although limited to twenty in number by an ordinance of 1660, these men effectively took the army under their charge: leading regiments to quarters, supervising garrisons, conducting reviews, overseeing the payment of troops, stamping out the abuse of the *passevolants*, investigating disputes, and guaranteeing discipline.[1]

Correspondence in the French war archives shows how ably these military commissioners fulfilled their tasks. When Benoist accompanied an army corps to its embarkation point at Toulon in 1660, he had to correspond with important figures, including the archbishop of Lyon, in order to complete the details for its passage—especially the duty of seeing that a sufficient number of boats were ready at the troops' arrival. At the same time, Le Tellier cautioned him "to take

---

1. Louis André, *Michel Le Tellier et Louvois* (Paris, 1942), p. 411. For Talon, see his joint commission with Honoré Courtin "pour travailler avec les comʳᵉˢ deputtez par le Roy Catholique a regler les limites des deux royaumes du côté des pays bas en consequence du traitté de paix signés le 7ᵉ Novʳᵉ 1659," 6 April 1660, A.G. A¹ 164, pp. 120r–25r.

good care that the troops live in order during their march so that no complaints are received." Other military commissioners were also busy. Colin prepared the barracks at Pignerol for the men to be garrisoned there. Charpentier readied the fortress of Guise and Ham for lodging several infantry companies; he had to make agreements with local entrepreneurs to furnish firewood and candles to the soldiers. Le Tellier approved an advance payment to the contractors—"observing only that you," he reminded Charpentier, "use the greatest economy possible in this." [2]

Undoubtedly, the commissioners' most important duty was to conduct reviews of the garrisoned troops. According to the series of letters between Le Tellier and commissioner Colin at Pignerol, on 19 November 1660 the secretary of war warned: "The king wishes you to make your review very carefully [*bien exacte*] and you are to see to it that those who make use of *passevolants* are chastised with the utmost severity. Payment [of troops] is to be made only for the effective force." One month later the secretary repeated his orders to Colin, remarking that there were now more *passevolants* in the regiment of Normandy than when the commissioner arrived at that garrison, as well as numerous native Piedmontese masquerading as French soldiers: "It appears to me that everything is in such poor condition that I fear the officers will feel the marks of the king's anger unless they change their conduct very soon." Again Le Tellier cautioned him that no *passevolants*, Piedmontese, or unequipped soldiers were to be permitted in the reviews; nor were the troops to be paid except on the basis of these reviews. The secretary noted that the officers who demanded extra sums to bring their companies up to strength were only using this as an excuse to get more money for the same number of men.

Evidently Colin could not satisfy the minister, for on 28 January 1661, Le Tellier wrote to the commandant at Pignerol announcing that Colin had been replaced by a commissioner named Amoresan. The secretary bluntly criticized the commander's actions: instead of aiding Colin in tracking down the *passevolants*, the unequipped soldiers, the Piedmontese, and the underaged youths who presented themselves for reviews, the chief officer had listened to the schemes of his officers. After Colin made a review at which 709 were present, the commander ordered another review, at which he found 774 men present. "I cannot hide from you," Le Tellier wrote, "that your conduct has not been approved; you have exceeded your power, for the

2. Le Tellier to Benoist, 10 December 1660, A.G. A¹ 163, fol. 331; to Colin, 19 November 1660, A.G. A¹ 163, fol. 252; to Charpentier, 21 January 1662, A.G. A¹ 172, fol. 91.

authority you have in the place [Pignerol] only goes to support the functions of the *commissaire* in the reviews he makes of the troops, and not to perform his function." And although Amoresan had several years' previous experience, the war secretary once again repeated the injunction he had given Colin: "pay only upon the basis of the effective force that you find at the reviews." [3]

Military commissioners handled the same variety of tasks as the army intendant. Le Tellier constantly demanded that they report everything of significance that occurred among the soldiers. On 2 November 1663, the war secretary wrote to Amoresan to render an exact account of the condition of a foundry established in the fortress at Pignerol by a person named Cron. "And to oblige him not to lose one moment of time, it would be good if you let him know that you have orders to observe him, and to write us exactly how he advances." No doubt this subtle hint spurred the foundryman to greater efforts, but it was no unusual demand of Le Tellier's. "You have given me no news," he admonished another commissioner, La Tournelle, "of the condition of the troops that are in garrison there." He should be more prompt in the future, the secretary warned. [4]

Like army intendants, military commissioners sometimes investigated misconduct among the soldiers, or at least provided information to the war department. When four army officers had fought among themselves, Le Tellier wrote to commissioner Aubert asking him to identify the persons involved, "for I do not believe that one can hide the truth from you in this respect as much as one tries." On another occasion, when the captains of the King's Regiment united in a body to complain about Le Tellier's orders and deputized one of their number to seek justice from the king, the war secretary sardonically wrote to *commissaire* Esmale: "I do not doubt that they will be very surprised when they see that His Majesty has ordered you to assemble the regiment." Esmale was to read before the soldiers a royal order dishonorably discharging the regiment's representative, the sieur de La Rivière, and instructing the provost to take prisoner all those who had conspired with La Rivière. The commissioner's action was to serve as an example, the secretary added, that those who bore the name of His Majesty's regiment owed much more obedience than those who did not have this honor. [5]

3. Le Tellier to Colin, 19 November and 31 December 1660, A.G. A¹ 163, fols. 252, 392; to M. de Bertonnière, 28 January 1661, A.G. A¹ 168, fol. 94; to Amoresan, 24 February 1661, A.G. A¹ 168, fol. 144.

4. Le Tellier to Amoresan, 2 November 1663, A.G. A¹ 181, fol. 3; to La Tournelle, 15 November 1660, A.G. A¹ 163, fol. 235.

5. Le Tellier to Aubert, 1 December 1661, A.G. A¹ 170, fol. 336; to Esmale, 29 June 1663, A.G. A¹ 178, fol. 422.

Often the military commissioners themselves got involved in disputes with officers. In such cases Le Tellier sought an intendant's judgment, clearly revealing that the intendants retained the superior rank. In 1663, for example, *commissaire* Deslandes quarreled with Captain Blesse of the Queen's Regiment over the exact number of men in his company. Although the provincial intendant of Roussillon, Macqueron, investigated the incident, he found no great wrong on the commissioner's part. Le Tellier disagreed. It was good that the commissioner had carried out his duties exactly, but he should exercise more caution so "that he never maltreats, either by word or deed, the officers of the troops." [6]

Since the days of Louis XIII, there had been a tendency to station these *commissaires* in the provinces as well as the armies. For this reason, the commissioners often substituted for provincial intendants in military affairs when the latter were incapacitated. In 1660 when François Villemontée was forced by illness to curtail his activities in the generality of Soissons, commissioner Etienne Carlier substituted for him. In this capacity Carlier paid the troops, published ordinances from the king requiring the regiments to be complete, and conducted reviews of troops. He even had other commissioners under his authority, although he never received a commission as a provincial intendant. [7]

While the military commissioners were busy occupying themselves with the problems of a peacetime army, Le Tellier used the time to instruct his own son and future army intendants. For many years the secretary had planned to hand his office over to François-Michel Le Tellier, the marquis of Louvois. For this purpose, he secured in 1655 the right of *survivance*—succession in the secretaryship—for his offspring. In 1660 the twenty-year-old Louvois entered the war department as an apprentice, obtaining a brevet on 24 February 1662 allowing him to exercise the charge of secretary of state in the absence or illness of his father. In 1664 Louvois acquired the right to sign documents in his father's presence. Gradually the young marquis learned the paperwork of the secretariat, first by writing acknowledgments, then by addressing letters to the *commissaires des guerres* and, later, letters to the intendants and generals.

In the beginning, these letters concerned only the details of administration—reviews of soldiers, their pay, matters of supply. After-

6. Le Tellier to Macqueron, 18 May 1663, A.G. A¹ 178, fol. 139.
7. Le Tellier to Carlier, 25 May, 3 September, and 16 October 1660, A.G. A¹ 162, fol. 192, and A¹ 163, fols. 14, 133. On 3 June 1661 he was included in a list of intendants to whom Le Tellier sent a letter concerning the payment of troops for the month of April, A.G. A¹ 169, fols. 22–23.

ward they took the form of ordering troop movements and announcing policy decisions. Louvois also opened the dispatches addressed to his father, and following the years 1663–64, those to himself. Louvois did not take over the war department either in 1662 or even by the time of the so-called War of Devolution in 1667. This period was his apprenticeship. But he assumed the major responsibility for the everyday business of the war department, although Le Tellier continued to advise and work with him until 1677, when the father became chancellor of France. One cannot precisely locate the time when the son acquired primary authority and the father took a secondary place; it is sufficient to conclude with Louis André that until 1677 Louis XIV had two secretaries of war who mutually supported each other.[8]

While Le Tellier introduced his son to the war department, he began to train a number of future army intendants to serve under Louvois. Although there exist no letters from Le Tellier openly announcing this policy, the conclusion is inescapable. In the past, Le Tellier had arranged the appointment of several close friends and relatives to intendancies: Pierre Goury, Louis Chauvelin, Charles Brèthe de Clermont, and Philibert Baussan. Already the secretary of war had built up a clientage system, and it might be expected that he would continue the practice.

The innovative element was the careful preparation these new men were given. Le Tellier appointed them to minor intendancies between 1660 and 1667, where they could get the feel of the job and acquire the experience necessary for the next war. Three key men were groomed during this period: Jacques Charuel, Etienne Carlier, and Louis Robert. Both Charuel and Carlier were recruits from the ranks of the *commissaires des guerres*, an indication of the growing importance of that office.

The first mention of Charuel in the war archives is an order in 1645 to the treasurer of war to pay several men, including "Jacques Charuel, controller in ordinary of war." A controller normally accompanied a *commissaire* in conducting reviews and kept the muster rolls of the troops. The controllers passed into obscurity, however, while their companions grew in significance. In 1650 Charuel was a *commissaire* ordered by the king to negotiate an agreement on contributions on the northern frontier. Various other references to Charuel as *commissaire des guerres* exist, but he seemed to drop from sight after 1651, at least in official documents. Possibly it was during this period that he served as secretary to Le Tellier, for genealogical records in the Bib-

8. André, *Michel Le Tellier et Louvois*, pp. 55, 66–68, 72, 143–46, 272–74, and passim.

liothèque Nationale mention this fact, although they do not give a date.[9]

Charuel's opportunity came in 1664 when the crown organized an expedition against Tunisian and Algerian corsairs along the coast of North Africa. A commission dated 6 June 1664 named Charuel as "intendant of our finances and of the police in our land army which will be joined to our naval force." The expedition was a disaster. A fleet of sixty-three vessels set sail from Toulon on 2 July 1664, joined a contingent from Malta, and arrived off the coast of Africa on 21 July. The naval force occupied the two coastal towns of Bougie and Djidjelli, but the commanding officers, the duke of Beaufort and the count de Gadagne, quarreled; supplies ran low; and the Moors, reinforced by native artillery and guerilla tactics, threatened to overpower the expeditionary force. The French began evacuation by the end of October, abandoning some thirty cannon and losing nearly two thousand men. Although the French war archives preserved no letters from Charuel describing his role, it was not an auspicious beginning for the young army intendant. Yet he would prove to be one of the war department's most capable agents.[10]

Another of Le Tellier's *commissaires des guerres*, Etienne Carlier, attracted the war secretary's attention, but little is known of his work as a military commissioner. In 1660 he substituted for Villemontée, intendant of Soissons. He seems to have remained as acting intendant during the early 1660s, for he is mentioned in a list of intendants who were sent orders concerning the payment of troops in June 1661. But he did not have the legal title of intendant, and other documents persist in addressing him as *commissaire des guerres* as late as March 1665. His real chance came in the summer of 1664, when he accompanied the French expedition to Erfurt as intendant of the troops under General Pradel. The crown had sent a military contingent to help its ally the archbishop of Mainz, Johann Philipp von Schönborn, regain con-

---

9. Order to Charles de Lancy, *trésorier général ordinaire de nos guerres,* 14 January 1645, A.G. A¹ 91, pt. 1, fol. 124, and the agreement (*traité*) made by Charuel, "commissaire ordonné par le Roy pour traitter des contributions des places frontieres et conquises," 23 October 1650, A.G. A¹ 122, fol. 392. For other mention of Charuel, see the instructions, 29 April 1651, regarding the levies that had been made for the payment of the *ustensile* to the garrison of Béthune, A.G. A¹ 126, pp. 152r–54r, and the instructions to regulate with the deputies of the king of Spain the contributions on the frontier of Picardy, Flanders, and Artois, 27 November 1651, A.G. A¹ 127, pp. 227r–30r. For Charuel's genealogical records, see B.N. *dossiers bleus* 171, family "Charuel," fols. 2–13.

10. Commission of Charuel, 6 April 1664, A.G. A¹ 184, fols. 329–30; Ernest Mercier, *Histoire de l'Afrique septentrionale (Berbérie) depuis les temps les plus reculés jusqu'à la conquête française (1830)* (Paris, 1888–91), 3:262–66; Camille Rousset, *Histoire de Louvois et de son administration politique et militaire* (Paris, 1862–63), 1: 78–81.

trol over the rebellious town of Erfurt. At the conclusion of the campaign, Carlier returned to his job as military commissioner for the garrisons of Hainaut. The following year, 1665, the government named him "intendant of our finances and police in the corps of troops that we are sending into Holland." Again he served under Pradel, this time to assist the Dutch against the marauding soldiers of the bishop of Münster.[11]

Louis Robert was the intendant closest to the heart of the secretary of state, for he was a relative of Le Tellier, who first employed Robert as a secretary, then sent him to work under Jacques Brachet, intendant of the army of Italy, although strictly speaking it does not seem that he held the office of *commissaire des guerres*. When Brachet died late in the summer of 1659, the minister informed Mazarin of the fact and suggested that Robert take his place. The cardinal replied: "It is fortunate that Robert finds himself there at this juncture, and, as he has knowledge of everything, he would be the most appropriate to take care of the troops and execute what has to be done. . . . You may therefore, the king approving, take the trouble to address the necessary orders to him, and I do not doubt but that all will pass very well, seeing by the letters of M. de Navailles that he applies himself there in good fashion." Although the end of the war in November meant the return of French troops, Robert acquired actual, albeit brief, experience as an army intendant.[12]

---

11. Letter of Le Tellier, 3 June 1661, including a list of intendants, A.G. A¹ 169, fols. 22–23. Regarding the expedition to Erfurt, see Georg Mentz, *Johann Philipp von Schönborn Kurfürst von Mainz Bischof von Würzburg und Worms, 1605–1673: Ein Beitrag zur Geschichte des siebzehnten Jahrhunderts* (Jena, 1896–99), 2:84–88. There are no documents regarding Carlier's intendancy preserved in the French war archives. Proof of his intendancy rests upon Georg Mentz and the genealogical records of the Bibliothèque Nationale, particularly the letters of confirmation of nobility for Carlier, July 1669, B.N. *cabinet d'Hozier* 78, family "Carlier," fol. 2, which mention Carlier's past service as "intendant de la police et des finances sur le corps de troupes que nous avons envoyé soubs le commandement du Sr de Pradel . . . a lexpedition d'Erfurt, Allemagne." Carlier then returned to his original position as *commissaire des guerres*, for there is a receipt signed by him, 6 April 1665: "nous Estienne Carlier con^er du Roy comm^re ordr des guerres a la police et conduite des troupes estant en garnison dans les places de Hainault," B.N. *pièces originales* 598, family "Le Carlier en Picardie," fol. 5. The entire volume of A¹ 198 in the French war archives is devoted to copies of correspondence regarding Carlier's intendancy and the expedition under Pradel to aid the Dutch. Louis André confused Etienne Carlier with his brother Pierre, the *commis* in the war department. The Carlier genealogy in the Bibliothèque Nationale is terribly confused, but for proof that Etienne was the army intendant, see the letters of confirmation of nobility cited above in this note—notwithstanding André's accounts, *Le Tellier et Louvois*, pp. 421–22, and *Michel Le Tellier et l'armée*, p. 645*n*1.

12. B.N. *dossiers bleus* 569, family "Robert (à Paris) Sr de Montault," nos. 2–3; Mazarin to Le Tellier, 15 September 1659, in Georges d'Avenel, ed., *Lettres du*

In 1660 Le Tellier advanced the career of his protégé once more, for Robert obtained the intendancy of the corps that the king sent to Crete. The secretary did not entrust the young and still inexperienced man with the full responsibility of an intendancy—he received a commission only as intendant of finance, not the fuller title of intendant of justice, police, and finance. Furthermore, Guillaume Millet de Jeurs accompanied him in the double position of commissary general of the army and president of the council of war. Millet's commission as commissary general contained many provisions normally found in that of a regular army intendant: "conduct the *môntres* and reviews," "see that the payments are made to the soldiers of our army, conforming to and on the basis of these reviews," "see that the bread rations are distributed to the sergeants and infantry," "watch that no abuses are committed," "take care to make our men of war live in good order, police, and discipline," "hear the complaints and grievances of our subjects and others on the excesses and violences that our men of war do to them, pursuing all crimes and disorders that may be committed . . . so that no crime of consequence goes unpunished." Robert, a mere intendant of finance, had little authority beside such a powerful companion. Millet de Jeurs cosigned with Robert, or signed separately in his absence, the orders of the general for the payment of troops and other army expenses.[13]

The war department did not keep the complete records from this expedition; only scattered references indicate that Robert rather perfunctorily filled his intendancy. Before the campaign opened, for example, he traveled to Lyon to purchase a thousand pikes, a thousand muskets, four thousand pairs of shoes, and two thousand shirts to supplement the soldiers' dress and armament. In Crete, Robert fell ill for part of the expedition. The rest of the time he spent trying to convince Millet de Jeurs not to leave the army and return home without orders from the crown. In addition, Robert experienced that common affliction of all intendants: shortage of money. While he expected the arrival of sixty thousand *livres* in September, it did not arrive until the beginning of 1661. Although Le Tellier assured him that the army would not suffer, the expedition continually lacked funds. In spring

cardinal Mazarin pendant son ministère, Collection de documents inédits sur l'histoire de France (Paris, 1906), 9:308–10.

13. Commission of Millet de Jeurs as "Com^re gñal en notre armée de terre qui sera commandée par notred. cousin" Prince Almerick, 4 April 1660, A.G. A¹ 164, pp. 109v–12v. His brevet as "président au conseil de guerre" is dated 3 April 1660, A.G. A¹ 164, pp. 108r–9v. Compared with the commission for a commissary general of supplies in the army of Catalonia in 1631, A.G. A¹ 14, fol. 25, it is apparent how extraordinary Millet de Jeurs's powers were.

1661 the government decided not to send any new reinforcements, but not wanting to betray its Venetian allies, it promised to keep the present troops there and furnish a subsidy. The expedition did not return home to the port of Toulon until the following March (1662).[14]

At Toulon, Robert learned valuable skills in disembarking troops, conducting reviews, and paying the soldiers. In cooperation with La Guette, the marine intendant at the port, he distributed newly purchased clothing to the soldiers who wished to remain in the service— a distribution that elicited the usual caution from the secretary of state not to give any clothing to the disbanded troops and to keep the affair absolutely secret until the last minute. In all, it was a rather commonplace intendancy, the type of rehearsal that served as experience for the future.[15]

A year later, 1663, Robert had the opportunity to serve as intendant of police and finance in the vanguard of troops sent to northern Italy to apply pressure on the papacy after the "insult" offered to French ambassador Créquy. The expedition offered little excitement for Robert. He occupied himself with obtaining lodging, rights of passage, and supplies from Italian officialdom, and his intendancy was brief. That winter the government enlarged the force, placed it under the command of Du Plessis Praslain, and replaced Robert by Honoré Courtin for an expected march on Rome. Robert had completed his task satisfactorily, however, and the government switched him to another assignment, the intendancy of finance in the corps sent to help the Empire fight the Turks.[16]

14. "Mémoire envoyé a M. Millet s'en allant a l'armée navalle," 3 April 1660, A.G. A¹ 162, fol. 1. Le Tellier to Robert, 12 November 1660, and 14 January, 11 February, 14 April, and 3 June 1661, A.G. A¹ 163, fol. 223; A¹ 168, fols. 52, 117, 264; A¹ 169, fol. 15. Le Tellier's attitude toward Robert's illness illustrates the secretary's personal concern for his relative's career. In two letters to Millet de Jeurs, 17 September and 22 October 1660, A.G. A¹ 163, fols. 51, 156, the minister attributed Robert's cure to the commissary general's attention, employing in each of the letters the common phrase that he thanked him with all his heart—*de tout mon coeur.* In a separate letter to Robert, 12 November 1660, A.G. A¹ 163, fol. 223, the secretary rejoiced with the intendant on his recovery: "Jay eu bien de la joye d'apprendre par vre. lre. que vous estez graces a dieu guerie de vre grande maladie." He then proceeded to lecture Robert on his health, warning him to preserve it from any setback because one must value it above all other things in the world, which were worthless if one could not enjoy them. Such solicitude is comparable to an overanxious mother's, but this concern was in complete accord with Le Tellier's character. He expected many things from his young relative and was determined not to be disappointed.

15. Le Tellier to Robert, 10 February and 28 March 1662, A.G. A¹ 172, fols. 183, 366; to M. de la Guette, 28 March 1662, A.G. A¹ 172, fol. 372.

16. The collection of letters relating to this expedition is copied in volume A¹ 182 in the war archives. Robert's commission is dated 26 September 1663, A.G. A¹ 182, pp. 233–36; an amplification of this commission to extend his authority to

HUNGARIAN EXPEDITION OF 1664

The role of army intendant Robert in an expedition in the 1660s illustrates in detail the process of apprenticeship and the development of the office since the days of the 1630s. The twenty-eight-year-old Robert, although knowing no German, showed extreme resourcefulness in overcoming the inherent difficulties of the campaign: the lack of supplies and their poor condition when available, the large number of sick in the army, the problems of finance and money exchange in a foreign land, and the difficulty of living amid the hostility of the imperial forces. Highly competent in matters of finance, he also exhibited successful qualities of leadership, particularly in delegating authority to his subordinates, the *commissaires des guerres*. One must not forget that this was Robert's fourth, although the most important, campaign. Experience combined with intelligence had produced a highly qualified administrator.

The French force that went to the aid of the Empire was tiny, numbering only four thousand infantry and two thousand cavalry, under the command of Count Jean de Coligny-Saligny. Coligny's instructions mentioned that he should assist the intendant in any way possible and act in concert with him regarding the expenses of the troops. But Le Tellier later wrote Robert that he had spoken to the general privately: "I told him plainly at his departure from court that you were my relative and had served the king at my side, that I had affection for you, and that it would give me pleasure if he were considerate of you." [17]

Robert's instructions were precise. As intendant of the troops stationed in northern Italy, he was to march through Venetian territory with twenty-six companies of cavalry to Marburg in present-day Yugoslavia to join the main body of French forces under Coligny. "The intention of His Majesty is," the instructions continued, "that Sieur Robert have a very special care to furnish the supplies needed by the troops," that is, in accordance with arrangements made by the French ambassador at Venice and by purchases in Germany according

the lands of the duke of Modena, 9 November 1663, A.G. A¹ 182, pp. 442–44; Courtin's commission is in the same volume, A¹ 182, pp. 873–76, undated (31 January 1664). For the background to the Italian situation, see André, *Le Tellier et Louvois*, pp. 115–32.

17. Coligny's instructions, 14 April 1664, A.G. A¹ 189, pp. 26v–41r, particularly p. 40v; Le Tellier to Robert, 23 October 1664, A.G. A¹ 189, pp. 242v–46r. Le Tellier had, however, mentioned in this letter that he expected the intendant to avoid a confrontation with Coligny: "et que je me rendois caution que vous [Robert] ne vous aplicqueriez a autre chose qu'a celles qui regardent vostre employ et qu'au surplus vous recherchieriez avec soin les ocasions de luy complaire."

to the official rate negotiated by Gravel, the French plenipotentiary to the German diet. To assist Robert in obtaining supplies, the king had contributed funds for the establishment of grain magazines for the use of the allied armies. To help him coordinate such supplies and to arrange lodgings, the instructions advised Robert to send an intelligent and capable person in advance to make arrangements for the army's arrival. For this purpose, the war department named a number of *commissaires des guerres* to accompany the army, although their disposition was left to the commanding officer and Robert, who "will employ them in the functions of their office as they think appropriate."

The crown set aside ten thousand *livres* for medical care. If Robert had not already acquired the necessary furnishings, drugs, and medications, his instructions warned, he should do so as soon as possible. The soldiers themselves would bear the cost of the hospital through deductions from their pay, supplemented by any revenue left over in the army after ordinary expenses. The king even considered the spiritual needs of his army, giving the intendant a fund of six thousand *livres* to maintain four Jesuit fathers with the corps. Finally, the last clause of Robert's instructions asked him to report frequently the news of what happened in the corps of troops and in the allied army. To facilitate this, the war department gave him a special code to use in writing home.[18] Robert's instructions were not very different from those given to previous intendants. The same prescriptions occur over and over: deduct the cost of the bread from the pay of the troops, pay the men on the basis of exact reviews, oversee all expenditures, maintain order and discipline, and tell the government about everything that happened on campaign.

Part of Robert's job was to arrange for the departure of French troops from Italy, namely from Mantua, Parma, and Modena. This entailed corresponding with the ministers of these governments to work out details of routes and supply depots. After investigating the situation in Mantua, Robert decided not to hire entrepreneurs, but to use the supply assistants, *commis des vivres*, to buy the necessary foodstuffs, hoping in this way to get better-quality goods at a lower cost. But the intendant judged it best to use local contractors in Venetian territory. Robert had the assistance of Pierre de Bonzi, the French ambassador, in working out the details of the routes, for the Venetians were suspicious of troops passing through their territory. At the same time, Robert wrote ahead to Gravel questioning the forage situation

18. This and the preceding paragraph are from Robert's instructions, 7 May 1664, A.G. A¹ 189, pp. 108v–17r. I did not find Robert's commission in the war archives.

in the Empire. Since he did not know the native weights and measures, he asked Gravel for information. Gravel assured him that there was an adequate quantity of supplies, but suggested that he send advance men to oversee their preparation and distribution. Consequently Robert sent a *commissaire*, Charles Camus du Clos, ahead "to enlighten himself on all things there and inform me on my arrival." [19]

By far the most difficult part of the preparations was the problem of monetary exchange. Robert puzzled over the matter of how the French government could exchange its money at the lowest rate. The intendant reported that of the one hundred fifty thousand *livres* he had received, it had cost six thousand *livres* to change them into Hungarian *sequins* at the current rate of six *livres* to one *sequin*. Robert thought this exchange rate was intolerable and remarked: "I inquired of various bankers to see if one can change Parisian money with less loss in this town, but I have not found anyone who would hear of undertaking it at a lower exchange rate." He had looked into the possibility of changing the money in small towns along the route, but found that one had to use the native coinage for all transactions. For this reason, he changed the money directly into Hungarian currency, although this meant a loss of ten *sols* per *pistole*. Robert was not sure if this loss would be borne by the king or the troops, "but I am persuaded that you would not want to treat the soldiers so rigorously in a foreign country." Louvois replied that the army would have to bear the loss for at least a month, until he concluded an agreement with a banker to deliver money to Marburg in *rixthalers* by weight and at the rate of two-thirds for one *pistole*. In the meantime, Louvois sent Robert letters of exchange for an additional three hundred thousand *livres*, plus letters of credit for the same amount drawn on both Vienna and Venice, warning the intendant to "guard them [the letters of credit] carefully and use them only in case your funds are exhausted before receipt of the money from the letters of exchange—*avant l'acquit de vos lettres de change*." [20]

By this time the army was on the march. "Our troops continue to advance with all the order and discipline possible," the intendant reported from Villach in Carinthia; "I have visited several of the *étapes*—supply depots—where they have passed without receiving the least

19. Letters of Robert, 12 and 19 May 1664, A.G. A¹ 190, pp. 36r–43r, 51r–56v. Copies of Robert's and Coligny's letters are preserved in volume A¹ 190 in the war archives. In the letters that do not indicate which of the two war secretaries was the addressee, I have merely cited them as letters of Robert.

20. Robert to Le Tellier, 13 June 1664, A.G. A¹ 190, pp. 87r–91r; to Louvois, 13 June 1664, A.G. A¹ 190, pp. 91r–93r; Louvois to Robert, 7, 21, and 28 June 1664, A.G. A¹ 189, pp. 138r–39r, 150rv, 152r–53r.

complaint, but on the contrary all sorts of praise. I then visited each brigade, one after the other, and, having observed them carefully, I found that they maintain themselves in complete numbers. For the most part, the horses are in good condition although there are some which are injured or lame as inevitably occurs on such a long route." Then, having inspected the troops and seeing that his presence was no longer needed, Robert traveled ahead to confer with the imperial commissioners about matters of supply. He learned through his conversations with the commissioners that everything was ready in Carinthia, but that two supply depots in Styria were so disorganized that he must go on ahead to work out arrangements. Robert found everything there in complete confusion; one of the imperial commissioners offered to furnish the supplies, but they were of such poor quality that Robert hesitated to pay him.[21]

Things were even worse when Robert arrived in Marburg. He was surprised to discover that the town was unprepared for the troops' arrival. He could not find an imperial commissioner to deal with, only a list of lodgings that turned out to be "very confused and poorly regulated." Grumbling that the people there were as stupid as the peasants of France, he gave orders on his own authority for suitable housing to be readied. The food situation was equally bad. The munitioneer with whom he had expected to deal had voided his contract, and Robert had to dicker anew with the entrepreneur at a higher price. The contractor hoped to get three *sols* four *deniers* for a twenty-four-ounce ration of bread. The outraged intendant reported that he would not think of paying more than the rest of the allies. But this left him in a quandary. Although he could not imagine the confederate princes paying this price for their bread, it certainly cost more than the price stipulated in his instructions: fifteen *deniers* a ration.[22]

Meanwhile, Robert set to work to try to find additional bread for the army. He decided that it would be advantageous if the troops could carry their own bread with them for five or six days; therefore he set out to buy all the horses and carts available for transport. He obtained

21. Letters of Robert, 20 June and 2 July 1664, A.G. A¹ 190, pp. 104v–7v, 127v–36v. Robert had, however, received some complaints about disorder in Venice and in the Empire—although it seems the latter did not actually involve French troops destined for the campaign in Hungary. But Robert remained optimistic in his letter of 2 July 1664: "nos troupes ont vescu avec plus d'ordre que n'auroient fait des capuchins s'ils avoient esté en aussy grand nombre."

22. Letters of Robert, 2 and 9 July 1664, A.G. A¹ 190, pp. 127v–36v, 151v–54v. There were twelve *deniers* to a *sol*, so three *sols* four *deniers* would equal forty *deniers*, much more than the fifteen *deniers* the crown deducted from the soldiers' pay. Bread was not furnished free to the troops, but set at a standard rate and deducted from their salaries.

nineteen carts and began a search for flour. After much difficulty, he found a four-day supply and loaded his carts. This foraging expedition met disaster—rains fell and the carts became engulfed in mud. In addition, scouts reported that the Turkish army was in the vicinity, and Gassion, the commander of the cavalry, decided to rejoin the main body of Coligny's army, leaving the convoy of flour as well as most of the army's baggage behind. Robert complained bitterly of the decision, blaming Gassion for the lack of bread: "The stubbornness that he displayed . . . caused the loss of everything. I do not want to exaggerate, but assuredly if someone else had commanded in his place, this misfortune would not have happened." Robert pessimistically continued: "I see only confusion, disorder, and especially dearth, not only of bread, but also of the munitions of war, which are lacking everywhere." [23]

The government did not condemn Gassion's actions, because the great victory at Saint Gotthard Pass wiped out any consideration of such mundane matters as the loss of a flour convoy. Robert wrote a narrative of the battle and sent it to court. In his account, the intendant praised the actions of his friend the comte de La Feuillade, to "whom all the generals here concede the honor of this victory without dispute." But the victory did not change things in the army. The battle resulted in some five hundred wounded, and Robert confessed: "It is certain that the wounded are not well cared for and the troops, especially the infantry, suffer much from the lack of bread." Coligny confirmed that affairs were going badly: "Monsieur Robert and I are embarrassed, not being able to find means of obtaining bread for the soldiers." He blamed the trouble on the Imperials, for whom the French had risked their lives but who refused to supply the allies. Also the roads were impassable, and the continual rain made the countryside one vast swamp. To add to the difficulty, Robert fell ill, but he recovered in a short while.[24]

Coligny reported that the troops were murmuring about the miserable living conditions. "All of our officers and soldiers are very discouraged by the bad treatment they receive here. . . . For two weeks a certain bad humor [*un esprit de chagrin*] has slipped among the troops, fomented by men who are listened to with a little too much complaisance, which is a great prejudice to His Majesty's service." A week

23. Two letters of Robert, last of July 1664, A.G. A¹ 190, pp. 164v–71v, 173v–76v.

24. Letter of Robert, 7 August 1664, A.G. A¹ 190, pp. 180r–82v; accompanying this letter is "Relation des deux combats qui se sont donnez en Allemagne par les François contre les Turcs les 27. juillet, et premier aoust 1664 envoyée par Mʳ Robert," A.G. A¹ 190, pp. 182v–86v. Letter of Coligny, 8 August 1664, A.G. A¹ 190, pp. 207r–10r. It should be noted that Coligny's letter to the secretaries, 6 August 1664, A.G. A¹ 190, pp. 186v–89v, minimized La Feuillade's role in the battle of Saint Gotthard.

later, Coligny again complained of the grumbling in the army: "Since the troops are joined together, a certain pestiferous and caballing spirit has formed, which not everyone works to extinguish as they should. There are those here who win over hearts, who are prejudicial to His Majesty's service, and who make a little too much parade of their credit near His Majesty." [25]

Coligny did not name the intriguers, but undoubtedly he referred to the comte de La Feuillade, who had so recently demonstrated his prowess at Saint Gotthard, for Coligny was jealous of his ability. Robert reported that the general's ill will grew daily, and naturally Coligny's suspicions included the intendant because of his friendship with La Feuillade. Consequently, relations between the general and the intendant, already strained by the tension of the terrible weather, the closeness of the enemy, and the difficulty in finding food, deteriorated rapidly in the month of August. On 18 August 1664, Robert complained that the Austrian high command made the French march without rest or consultation. The intendant told Coligny that he could not take the proper measures to find bread unless he knew the army's plans. After a council of war, the French decided to send a delegate, La Feuillade, to remonstrate with the imperial commander Montecuccoli that the army could not march any farther unless the Imperials promised to provide bread. Montecuccoli told them that the route was well supplied, but Robert wished for a stronger assurance, including the desire that the French be consulted on all future imperial plans. Coligny did not want to press the issue, and Robert appealed to Le Tellier in private: "I remit the rest to what Monsieur du Clos will tell you by mouth," for when Camus du Clos traveled to court to recount the victory at Saint Gotthard, he also carried a personal message between Robert and Le Tellier.[26]

A few days later the disagreement broke into the open. When Robert

25. Letters of Coligny, 11 and 19 August 1664, A.G. A¹ 190, pp. 211r–13v, 225v–26v. Was there really a cabal against Coligny? True, he did not get along with Robert and La Feuillade, but I found no evidence that they conspired against him. Camille Rousset, however, thought that the two did try to take the glory of the victory away from Coligny; see his comments in *Histoire de Louvois,* 1:58—"Dans sa propre armée, il y avait un parti qui lui contestait son triomphe, pour l'attribuer à l'un de ses inférieurs. Parce que le comte de La Feuillade avait vaillamment payé de sa personne, parce qu'il avait exécuté, mieux que d'autres peut-être, avec plus d'intelligence et d'entrain, les ordres de son chef, il avait de nombreux amis qui faisaient de lui le héros de l'action, le vrai, le seul vainqueur des infidèles. C'est ainsi que l'intendant de l'armée, M. Robert, en avait écrit à la cour et à M. de Gravel, et il affectait un tel enthousiasme pour M. de La Feuillade qu'il se fit rappeler au sentiment de son devoir, ou plutôt de son intérêt, par Le Tellier, son protecteur et parent."

26. Letters of Robert, 27 and 18 August 1664, A.G. A¹ 190, pp. 253r–54v, 220r–23v.

heard that Montecuccoli, without consulting his allies, had sent the French instructions to depart for the Raab River by 29 August, he replied that this was impossible. Where would the army get enough bread? Although the Imperials said the French could obtain sufficient bread for the ten-day march from Ödemburg, Robert replied that it was impossible for the infantry to carry that much bread with them and that the town was three hours off the main route, which would create a further delay. Coligny answered that he would do as Montecuccoli desired. At this, Robert retorted that he thought it was better to save the troops than to obey the order. Coligny sneered that Robert could make any arrangements he wanted to with Montecuccoli, but that he was not going to delay the departure one minute. The intendant wanted a confrontation with the Austrians, but the general replied that he was not going to be the scapegoat in a quarrel between the French and their allies. The intendant did not reply to this accusation, and the argument ended. But when Robert wrote to the war department, he reported Coligny's innuendos to the secretaries' attentive ears, adding: "This is the first time that he [Coligny] has brought his feelings toward me out into the open, feelings that he has made only too apparent by his conduct." Justifying his actions, Robert explained that it was an intendant's duty to tell the general that the army could not survive along the Raab River unless the Imperials guaranteed the bread supply.[27]

The two secretaries wrote prompt replies. Louvois wanted an immediate explanation of the disorder in the army, but reminded Robert that he must get along with Coligny and avoid further confrontation with him. Instead, he could satisfy his anger by writing long letters to the secretaries. Le Tellier used much stronger language. In a formal letter to the intendant lacking his customary salutation, "monsieur my cousin," he told him that he had been wrong in challenging Coligny. "I feel obliged to say to Sieur Robert once and for all that he is charged by his office to represent to Monsieur de Coligny the impossibility of furnishing bread to the troops. . . . If Coligny judges that the inconvenience, which he has reported, is not one to stop the army or slow its march, Monsieur Robert must not insist upon it." The secretary continued: "The authority to command resides in the person of Monsieur de Coligny; the intendant is placed near him only to execute the things with necessary order and to warn the king if they are abused. Monsieur Robert must do this to the letter without ever interfering."[28]

27. Letter of Robert, 27 August 1664, A.G. A^1 190, pp. 254v–57v.
28. Louvois to Robert, 19 September 1664, A.G. A^1 189, pp. 201r–3v; Le Tellier to Robert, 27 September 1664, A.G. A^1 189, pp. 208v–13v.

It was true, Robert responded, that he had attempted to persuade Coligny to change his mind and that the general had become angry, "but I always spoke to him with the respect that I owe him, and when he lost his temper I was afraid of embittering him. You will note that he, too, recognized that I had not given him cause to be used in this manner, for that same evening he apologized." Since that time, the intendant assured Le Tellier, he had contented himself with executing Coligny's orders promptly, and the general seemed satisfied with his conduct. "For the rest, I keep myself from attempting to do anything on my own. When I am in the army, I do nothing at all without consulting him first, and when I am elsewhere I render him an account of my actions in writing." [29]

The secretaries, father and son, were quick to reassure Robert. Louvois explained that the intendant need not be upset, for the letter had been written out of "pure amity" so that Robert would act with caution and "obtain from your services the fruit that they merit." Le Tellier showed greater concern. If the secretary had made observations on his conduct, it was only to warn him so that he would not "fall into inconveniences." "Do not be alarmed because of what I have written you," he continued. "I only did it out of friendship for you and for your welfare. Profit from it, so that when Monsieur de Coligny returns to His Majesty, he will testify to your prudent conduct and the utility of your services, so that the good reputation [*la bonne odeur*] that he spreads about you will facilitate ways to procure employment for yourself when the next occasion presents itself." [30]

Coligny was not stupid. He remembered the words of Le Tellier at the beginning of the expedition. He knew that Robert was the secretary's relative and that the latter cherished him. Never did he openly criticize Robert's conduct; he even complimented him from time to time. But he continued to harbor resentment against this civilian upstart. Many years later, Coligny recounted in his memoirs that Le Tellier had not been satisfied with him on his return from the Hungarian expedition because of the intendant. He bitterly remarked: "I was confirmed in my opinion, that when one is not a creature and a toady [*créature et très humble valet*] of ministers, it is a waste of time to attach oneself to the service of kings." The incident hurt Coligny more than anyone, for the intendant continued to rise in Le Tellier's favor while Coligny never received an important command again, not

29. Letter of Robert, undated but probably 10 October 1664, A.G. A¹ 190, pp. 353r–55r.
30. Louvois to Robert, 31 October 1664, A.G. A¹ 189, pp. 254v–55r; Le Tellier to Robert, 23 October 1664, A.G. A¹ 189, pp. 242v–46r.

even in the next war, the War of Devolution. It was a lesson for the generals: one did not advance one's career by feuding with intendants.[31]

It would be unfair to conclude that Robert poisoned Le Tellier's mind against the general. Coligny's disgrace had deeper roots. The general's reputation suffered because of his incompetence: his inability to get along with civilian control or even his own officers such as La Feuillade, his refusal to assert himself in imperial councils, and his incapacity in controlling his own army. The crown was upset especially that Coligny did not remonstrate strongly to the Austrians about their conduct in not supplying bread for the army or admitting the French to their councils. Even before the incident, the king had written to Coligny to inform the Imperials that unless they furnished bread at a reasonable price, the French would search for their own supplies. Louvois absolved the intendant of all blame, and stressed: "You must contribute as much as you can on your part to prevent and chastise the disorders and to encourage Monsieur de Coligny to bring the necessary remedies to bear; but if he does not apply himself as he should, you must not be uneasy; your solicitude suffices to acquit you of your duty and it will be for him to answer to the king." [32]

Unlike Coligny, Robert fulfilled his tasks to Le Tellier's satisfaction, particularly in the two essential areas of caring for the sick and feeding the army. Immediately after hearing of the battle of Saint Gotthard, the government inquired about the condition of the wounded, directing the intendant "to take all possible care to assist the soldiers and to relieve them in their distress." Robert estimated that in mid-August there were five to six hundred men who were unable to fight. He tried to place them in Ödemburg or Graz, but Montecuccoli refused to give an order to receive them. "However, I hope that one will receive them there," Robert wrote, "without his [Montecuccoli's] mediation." [33]

As the French army marched along the Raab River, the care of the sick became increasingly difficult. The problem remained one of locat-

31. Comte Jean de Coligny-Saligny, *Mémoires du comte de Coligny-Saligny*, ed. Monmerqué, Société de l'histoire de France (Paris, 1841), p. 102. Monmerqué notes, p. 102*n*1, that Bussy-Rabutin also recounted in his memoirs: "Il [Coligny] s'étoit brouillé, dit-il, avec un intendant d'armée que M. Le Tellier lui avoit fort recommendé." Although Coligny did not have an army command in the War of Devolution, he did serve briefly, and Monmerqué quotes one of his letters to his friend Bussy-Rabutin (p. xxxi): "j'ai le plaisir, à l'âge de quarante-neuf ans, de faire le métier de volontaire." Shortly thereafter he retired.

32. Louvois to Robert, 26 August and 19 September 1664, A.G. A¹ 189, pp. 187v–91r, 201r–3v.

33. Louvois to Robert, 26 August 1664, A.G. A¹ 189, pp. 187v–91r; Robert to Le Tellier, 23 August 1664, A.G. A¹ 190, pp. 233v–36v.

ing a hospital. Robert first thought of taking the wounded to Komorn and establishing medical facilities there, but then the Turks moved near Neutra, and Montecuccoli gave the French orders to cross the Danube and take up quarters in the region of the Waag River. As soon as he heard about the change in orders, the intendant thought of Pressburg and traveled there to see if he could establish a hospital. The town gave him several buildings and he rented others, even constructing temporary wooden barracks. The doctors of Pressburg satisfied the intendant's standards and he bought medicine, while the sick and wounded ate good bread and beef bouillon to strengthen themselves. Finally, Robert sent a man to Vienna to rent enough carts to transport the medical supplies, bedding, and linen left at Radkersburg to Pressburg, where they were needed.[34]

Meanwhile, Robert stayed at Pressburg to provide for the care of the sick and to find bread for the troops. He left an assistant, Camus du Clos, with the army to run things in his absence. The number of sick increased at Pressburg until Robert reported at the end of September that there were seven to eight hundred soldiers in the hospital, plus another one hundred to one hundred fifty valets. The supplies he ordered from Radkersburg did not arrive soon enough, which meant that he had to buy additional medical supplies in Pressburg that sold for "their weight in gold." Exhausted by his efforts, the intendant moaned: "I can speak only of two things, the bread and the hospital. They are the only things I apply myself to at present. . . . I am, myself, the munitioneer of the army and the director of the hospital." Everything else he had left to Camus du Clos. It is a measure of Robert's talents as an administrator that he knew how to delegate authority, for he had reserved the most important aspects of his employ for himself, distributing the lesser tasks to his subordinates.[35]

The other army officers testified to the efficacy of Robert's efforts. La Feuillade, as might be expected, was effusive in his praise: "Our intendant saves the army by his hospital and his diligence for the bread. . . . You do not know how to cherish him too much." Even Coligny joined in praising him: "We have nearly six hundred sick in our hospital at Pressburg; Monsieur Robert walks among them and revives them as much as he can, if not by his science at least by his care, like another Aesculapius. . . . He obtained lodging, built barracks, and he has done all that anyone could think of." [36]

34. Letters of Robert, 9 and 16 September 1664, A.G. A¹ 190, pp. 277r–87r, 292v–303r.
35. Letter of Robert, 29 September 1664, A.G. A¹ 190, pp. 316v–19v.
36. La Feuillade to Louvois, 24 September 1664, and Coligny's memorandum of the same date, A.G. A¹ 190, pp. 315rv, 311r–15r.

At the end of October, the army learned that it was returning home. Naturally the intendant had to make some provision for those who were too sick to travel, for the king did not want to leave anyone at Pressburg, because there would be no one to take care of them after the army's departure. Robert rented carts to transport the sick, reporting that the fear of being abandoned had given legs to many. Of the hundred and thirty or forty who remained incapable of marching, the carts carried a hundred, and the rest were left at Pressburg under the care of a *commissaire des guerres*. Louvois cautioned the intendant that it would be a good idea to leave his brother, the sieur de Loupigny, to watch the commissioner's conduct and to countersign all necessary expenses, although Robert hoped to close the hospital within a month after the departure of the troops.[37]

After the battle of Saint Gotthard, the problem of feeding the army increased. Despite imperial assurances that the French army would not lack bread, events soon confirmed Robert's suspicions. When the imperial commissioners did furnish flour to the French, the intendant reported, "it is so bad that I never use it except under protest." The flour contained a high percentage of waste material: a hundred pounds of it made only sixty rations of bread, although the French had paid enough to bake a hundred rations. The bread supplied by the imperial commissioners—when they did supply it, Robert sarcastically remarked —was of poor quality, composed mostly of rye and buckwheat, and the soldiers cried aloud when they saw it, while the officers threw away half the bread to be distributed.[38]

The intendant suspected that the poor quality played a role in the large number of invalids in the camp. This was why he preferred to buy grain elsewhere when he had the opportunity and not get it from the imperial commissioners. During the first week in September, for example, he went to Vienna to find an alternate supply. The trip was a failure, but it offers a glimpse of Robert's methods. Upon arriving in the capital, he searched out three "of the best and richest bakers" in Vienna and talked with them about the possibility of furnishing six thousand rations of bread per day. Robert refused to conclude the deal

37. Letter of Robert, 22 October 1664, A.G. A¹ 190, pp. 369r–71r; Louvois to Robert, 18 October 1664, A.G. A¹ 189, pp. 234r–37v. Louvois did suggest in this letter that some *commissaires des guerres* were better fitted than others to the task of staying with the sick troops: "Il est apropos que le comm^re de Vreuin demeure aupres d'eux pour ordonner du payement qu'a son deffaut, soit par maladie ou autrement, l'on y fasse rester le comm^re La Bussiere, mais quelque chose qui arrive le comm^re Le Vacher ne doibt point avoir cet employ. La connoissance que j'ay de leurs talents m'oblige a luy faire cette gradiation."

38. Letters of Robert, 1 and 9 September 1664, A.G. A¹ 190, pp. 262v–67r, 277r–87r.

until he had tested the quality of the bread, so he asked each baker to present a sample loaf in order to select, by competition, the best offer. The next day Robert waited for the samples, but none arrived. At first the intendant thought the contractors wanted more money, but he soon discovered there was a conspiracy in the city to prevent the sale of bread to the French. The imperial commissioners were involved, for they wanted to maintain a monopoly over supply in order to make a profit on the poor flour they furnished to the army. "I imagine," Robert added, "that it was made without the knowledge of the emperor, but by the singular avarice of some of his ministers, because self-interest reigns in this court more than in any other." The intendant overcame the conspiracy by buying the grain in small quantities from a number of different dealers and baking it into bread.[39]

Robert showed his resourcefulness in another way. Gravel reported from Germany that there were magazines of flour at Hainburg, Linz, and Ratisbon that could be used by the troops because the French king together with the allies had furnished money for their establishment. Robert decided to do a little detective work and sent an agent to Hainburg to investigate. The assistant found the magazine, but the guard refused to issue any flour without an order from the imperial commissary general. Robert pleaded with an Austrian officer, the count d'Holac, to give him special permission and sent Le Vasseur, his *commis des vivres*, to get the flour. After some difficulty, the guard at last succumbed to Robert's pleading, and the intendant obtained enough flour to make thirty thousand rations of bread. No one demanded payment, at least not until the intendant tried to draw another thirty thousand rations. The guard refused to allow the French to draw out any more flour unless they paid for it, but Robert angrily refused, complaining that the flour was not of the best quality and that the king had already contributed to the allies for its purchase. Despite this setback, the intendant did very well indeed, for he proudly wrote to the war department: "Since we have been near the Danube, I have not allowed our troops to lack bread, and I hope to take such good measures that they will not lack in the future." [40]

The emperor's armistice with the Turks and the subsequent French decision to bring its troops home reduced the burden of subsistence, but they raised new problems. Gravel busied himself by negotiating the

---

39. Letters of Robert, 9 and 16 September 1664, A.G. A¹ 190, pp. 277r–87r, 292v–303r.

40. Letters of Robert, 1, 9, 16, and 23 September and 6 October 1664, A.G. A¹ 190, pp. 262v–67r, 277r–87r, 292v–303r, 308v–11r, 334r–39v. This magazine at Hainburg had three different spellings in Robert's letters—Haimbourg, Hambourg, and Harbourg—but it is Hainburg, a town in lower Austria on the Danube, to which he is referring.

homeward routes and the furnishing of supply depots along the way. With foresight, Robert suggested that Gravel arrange for a quantity of wood at each rest stop, because the weather would be cold that winter. Without logs, the intendant warned, the soldiers would tear down neighboring houses for fuel.[41]

The departure of the troops raised another problem. Accounts had to be closed in the Empire, where the intendant hurried to pay the debts for supplies and transport. But, he complained, the Germans were the slowest people in the world; consequently it would be impossible to close the accounts in a short time. After Coligny and the troops left, the intendant decided not to stay his departure any longer. He promised the army's creditors that they could make an appointment with him at Prague or else they could settle any unfinished business with the chevalier de Grémonville, the French envoy in Vienna. To add to Robert's troubles, Louvois kept demanding expense accounts for the months of September and October, which the intendant feverishly worked to complete by 2 November 1664.[42]

The return trip was uneventful. An imperial commissioner accompanied the army on its way through Moravia and Bohemia and furnished supplies from the depots along the way. No one even asked the French to pay. After the soldiers reached Philippsburg, the French fortress on the Rhine, they disbanded for winter quarters in Alsace and Lorraine, where they came under the jurisdiction of two provincial intendants, Jean-Paul de Choisy and Charles Colbert. Robert sent the intendants a list of the troops according to the last review, and even traveled ahead to confer with Choisy, but he left all the arrangements in their hands. Meanwhile, he journeyed to Metz, where he persuaded the city lazaretto to care for the sick who had been transported there by carts. Then the intendant drew up the final accounts of the expedition and sent copies to the war department, before departing during the middle of January 1665 to render an oral report to the war secretaries.[43]

41. Letters of Robert, 6 and 12 October 1664, A.G. A¹ 190, pp. 334r–39v, 355r–57v.

42. Letters of Robert, 22 and 30 October, 2 November 1664, A.G. A¹ 190, pp. 369r–71r, 383v–84v, 387v–89v; Louvois to Robert, 18 and 24 October 1664, A.G. A¹ 189, pp. 234r–37v, 247r–48v. For Robert's final accounts, see "Mémoire de la despense faite pendant le mois d'octobre pour la subsistance des troupes du Roy qui sont en hongrie soubs le commandement de M le comte de Coligny," A.G. A¹ 190, pp. 389r–93v.

43. Letters of Robert, 12 November 1664 and 3, 6, and 10 January 1665, A.G. A¹ 190, pp. 399r–400r, 432v–33v, 436v–38r, 440v–41v. The French and the Imperials negotiated the route homeward through Moravia and Bohemia. I am not certain why this route was chosen, but see the letter of Robert, 22 October 1664, A.G. A¹ 190, pp. 369r–71r, which mentions that while this route was longer, it was also

On the way home, the army was torn by internal dissension. Coligny fought with the commissioners and then quarreled with the intendant. The general was a man who believed that civilians, both the intendant and his assistants, should submit to his wishes. Trouble began when the captains brought Coligny lists of vacancies in the officer corps, and he filled them on his own volition. The *commissaires des guerres* refused to accept the new officers without special orders from the king. "The truth is," Coligny complained to Louvois, "the commissioners are a little too punctilious and difficult in certain things and too easy in others. Unless one makes a great submission, one can obtain nothing from them." [44]

Coligny's views solidified following an argument with La Bussière over this same incident. During a verbal confrontation, Coligny had put his hand to his sword, whereupon La Feuillade, Podwitz, and other officers intervened. Robert placed the blame on La Bussière, who "drew it upon himself by his senseless words and his complete lack of respect." Le Tellier immediately wrote to the general expressing his surprise at the commissioner's conduct, but affirming that the *commissaires des guerres* were correct when they stated that no officer could be established in any charge without a brevet signed by the king. [45]

Shortly after this episode, an argument broke out between another military commissioner, Le Vacher, and a valet of Gassion, the commander of the cavalry. Le Vacher struck the man, and Gassion took his servant's side and complained to Coligny, although, according to Robert, "assuredly Monsieur Le Vacher had all the right on his side." Robert issued an order to seize the valet and sent a provost to arrest him at Gassion's lodging. Of course the man was not there, but Gassion felt himself insulted by having to submit to the ignominy of a search. Le Vacher complained that Coligny refused him justice, which the general indignantly denied: "When the commissioners have need of my support, they have not lacked it, but they rely upon themselves well enough and are so proud that the earth is not capable of supporting them." [46]

easier and desired by Coligny. Perhaps the adequacy of forage in these regions also helped determine the route.

44. Letter of Coligny, 10 December 1664, A.G. A¹ 190, pp. 417r–19r.

45. Letter of Coligny, 20 December 1664, A.G. A¹ 190, pp. 422v–24v; letter of Robert, 21 December 1664, A.G. A¹ 190, pp. 424v–26r. Le Tellier to Coligny, 30 December 1664, A.G. A¹ 189, pp. 286r–88r, in which letter Le Tellier did try to moderate Coligny's temper by adding his own opinion of La Bussière: "il m'avoit jusqu'icy parû un garçon sage, et des plus habiles de ceux de sa profession."

46. Letter of Robert, 21 December 1664, A.G. A¹ 190, pp. 424v–26r; letter of Coligny, 4 January 1665, A.G. A¹ 190, pp. 433v–36r. According to Coligny, not

After these disputes with Robert's subordinates, it is not surprising that the bitterness between Coligny and Robert once again broke into open hostility. Although the general always spoke politely of the intendant's efforts, he quietly sulked over Robert's behavior and found little ways to vent his spite. On 28 November 1664, Robert complained that he could not give an exact account of the march, because "I know only what everyone else knows; I do not interfere either with the march, or the routes, or the rest stops, or the furnishing of supplies, or the regulating of their payment. Monsieur Terruel handles this, about which I will have the honor of speaking with you in Paris. Let me content myself, however, by saying that we march with much confusion and disorder." Robert was justified in complaining that his functions as intendant had been taken over by one of Coligny's appointees, and Louvois promptly assured him that if he had known sooner, he would have put a stop to this practice, but as it was, the army was almost home and a letter would only serve to embroil the intendant further with Coligny—"from which you [Robert] would draw no advantage." [47]

About this time, Robert had a second disagreement with Coligny regarding an intendant's privileges. Precedence was a very important thing in the seventeenth century, battles were often fought over it, and Robert, as others in that century, was keenly aware of the importance of rank, though he professed otherwise: "I have no care but the service. I never attach to myself any privileges or rank, or any similar thing, which I count for a mere bagatelle." But, he insisted, if he passed the incident over in silence, "all the intendants can reproach me for . . . having poorly supported the rights and prerogatives of the charge that I have the honor of exercising." [48]

His complaint stemmed from the fact that traditionally the army intendant had the right to set up his tent in the position of honor beside the commanding general's. On the march homeward, the quartermaster sergeant marked out Robert's lodging next to Coligny's, but then Gassion came along and protested that he should be lodged in the place of honor beside Coligny. Gassion immediately complained to Coligny, who was standing nearby. The general called the quartermaster ser-

only had Le Vacher struck the man, but the valet "ayant esté outrageusement battu par les comm^res Le Vacher et La Bussiere qui s'estoient mis tous deux sur luy."

47. Robert to Louvois, 28 November 1664, A.G. A¹ 190, pp. 410r–13r; Louvois to Robert, 20 December 1664, A.G. A¹ 189, pp. 281r–83v.

48. Robert to Louvois, 28 November 1664, A.G. A¹ 190, pp. 410r–13r. For another example of a quarrel over precedence, this time between the Hôtel de Ville and the Châtelet in Paris, see Andrew Trout, "The Proclamation of the Treaty of Nijmegen," *French Historical Studies* 5, no. 4 (1968): 477–81.

geant, and according to Robert, "bullied and pushed him by words, as much as a man can be pushed, and even raised his cane to strike him because he had lodged me before Monsieur de Gassion." The sergeant changed the lodgings, and a storm of protest broke out, not only from Robert but also from his friends, La Feuillade and Podwitz, who knew that Robert's rank entitled him to the place of honor beside the commanding general.[49]

Robert complained to Louvois, but he received only the personal satisfaction of knowing that he had upheld his dignity. Louvois again said that he would have written to Coligny ordering Robert to take precedence over Gassion, commander of the cavalry, except that the letter would do little good at that late date. Instead, he urged the intendant to forget his scruples and to take extra care not to offend the general in the last days of the expedition.[50]

Coligny had a reckoning upon his return to France. In the general's own words, "I arrived at court in the month of January, 1665. I immediately visited Monsieur Le Tellier, who was not pleased with me because of several quarrels that I had with one named Robert, relative of the Sieur Le Tellier." The secretary was not pleased with Coligny because the crown preferred more compliant men who would work hand in hand with the civilian branch of the service, rather than an old-fashioned officer who thought the army was the independent fiefdom of aristocratic generals.[51]

Le Tellier and Louvois remained satisfied with their young relative. Robert had accomplished the difficult task of feeding the army and caring for the sick with talent and resourcefulness. Full of initiative, yet knowing how to delegate responsibility, Robert represents the ideal intendant of the period. Over and over again, particularly during the next war, the War of Devolution, the government utilized his services.

49. Robert to Louvois, 28 November 1664, A.G. A¹ 190, pp. 410r–13v.
50. Louvois to Robert, 20 December 1664, A.G. A¹ 189, pp. 281r–83v.
51. Coligny, *Mémoires*, p. 102.

VII

The War of Devolution
1667-68

Le Tellier's apprenticeship program would prove advantageous to the government in the coming war. His son, Louvois, had gained a working knowledge of the war department, and he had trained three army intendants as Louvois's assistants. Louis XIV also was anxious to fulfill what he regarded as his destiny. After a year of polished reviews in 1666, the king decided that the time was at hand to test his war machine. Taking advantage of a little-known legalistic argument to justify his wife's claim to the Spanish Netherlands—the right of daughters of a first marriage to share in an inheritance with male offspring of a subsequent union— Louis XIV began the War of Devolution. The resulting campaign was Louvois's opportunity. For the first time he exercised real responsibility, while his anxious father watched from the sidelines ready to correct mistakes.[1]

In mid-May 1667, the war began. The main body of troops, under the command of Turenne and in the presence of the king, operated between the Lys and Meuse rivers. Two smaller armies protected the flanks, that of Marshal Antoine d'Aumont in northern Flanders, and a similar force under the marquis de Créquy in Luxembourg. The French advance can only be described as a triumphal march. The conquest began with the occupation of Armentières on 24 May and ended with the taking of Alost on 12 September 1667. During the summer campaign the French captured Audenarde, Binche, Charleroi, Courtrai, Douai, and Tournai.[2]

The war department named three intendants for these armies. Louis

---

1. For "the year of reviews" (1666) and the preparations for the war, see Louis André, *Michel Le Tellier et Louvois* (Paris, 1942), pp. 157–65.
2. Camille Rousset, *Histoire de Louvois et de son administration politique et militaire* (Paris, 1862–63), 1:102–12.

Robert served with the troops under the command of Aumont, and Etienne Carlier with Créquy. The intendancy of the main army under Turenne remains obscure; during the time that the king and Louvois were present, there is no indication of an intendant. After their return to court, correspondence indicates that a master of requests, Charles Colbert de Croissy, served as army intendant. The date of Croissy's appointment is unknown, but it is possible that the government followed the tradition described in Bassompierre's memoir whereby the chancellor or secretary of state performed the role of intendant near the king's person. After the king's departure, the intendancy could be filled by a lesser person.[3]

When Louvois and the king were with Turenne's army during the main part of the campaign, there was no need for an exchange of letters between the secretary and the army, but Robert's and Carlier's correspondence gives a glimpse, albeit fragmentary, into their activities. Etienne Carlier's intendancy was the less active of the two. He arrived in the army in mid-May and remained on the Luxembourg frontier during much of the campaign. On 2 June 1667, he reported that Créquy had withdrawn large amounts of powder and munitions from the citadel of Metz, and Carlier promised to transport these supplies to the army. The intendant also concerned himself with the army's finances: he questioned the payment of five hundred *livres* in wages to the brigadiers while the commandant of the cavalry received only four hundred. Thinking a mistake had been made, he asked Louvois for further instructions. The answer, like so many others, is unknown, but this illustrates the intendant's constant attention to detail.[4]

3. The commissions of the three intendants are not preserved in the war archives, but there are scattered references to their titles in the correspondence. The bound *recueil* A¹ 209 of the letters written to Louvois in the first part of 1667 gives Robert the title of "intendant de l'armée de M. le Mar.ᵃˡ Daumont," fol. 29. There is a letter from Louvois to Carlier, 28 April 1667, announcing his army intendancy, A.G. A¹ 106, fol. 137. Jacques Charuel mentions "Mʳ Colbert, intendant de l'armée," in a letter to Louvois, 12–13 October 1667, A.G. A¹ 209, fol. 275. Charles Dreyss, ed., *Mémoires de Louis XIV pour l'instruction du dauphin* (Paris, 1860), 2:170*n*1, identifies him as Colbert de Croissy, as does Jean Bérenger in "Charles Colbert, marquis de Croissy," in Roland Mousnier, ed., *Le Conseil du roi de Louis XII à la Révolution*, Publication de la Faculté des lettres et sciences humaines de Paris-Sorbonne, series "Recherches," tome 56 (Paris, 1970), p. 159. Bérenger calls Colbert de Croissy's military intendancy "une mission de confiance." For the so-called Bassompierre memoir, see "En quoy consiste la charge et fonction d'intendant de justice et finances dans armée [*sic*]," B.N. *cinq-cents Colbert* 499, fol. 99.

4. Carlier to Louvois, 2 June 1667, A.G. A¹ 209, fol. 26. There are only selected letters recopied in *recueil* A¹ 209. Lack of complete correspondence prevents knowing the result of most of the incidents reported in this chapter.

Furthermore, Carlier complained, the treasury officials had not yet joined the army; they were at Metz awaiting money from Paris. Since it had not arrived by the middle of June, the soldiers began to grumble. The impatient troops took matters into their own hands and demanded money from their officers. The officers made little effort to stop their cries, and the army verged upon mutiny. Only the quick thinking of a lieutenant colonel of the regiment of Saint-Vallier saved the situation: he ordered the immediate arrest of one of the ringleaders and hanged him the following day, thus ending the disorder. Although Carlier as intendant of the army was supposedly charged with maintaining justice and policing the troops, he played no part in suppressing the mutiny. He contented himself with reporting the details to Louvois, merely remarking. "This punishment, to tell the truth, prevented the continuation of the disorder." Carlier's background was in military administration, not in law.[5]

Discipline remained a problem in Créquy's army. Near Betzdorf on 1 July 1667, Carlier reported that desertion had severely diminished the infantry because of the ease with which soldiers could slip across the border into the Spanish Netherlands. There the enemy promised them a passport and thirty *sols* for joining its army assembling at Liège. "As we are in a forested area," the intendant wrote, "it is very difficult to prevent this disorder. Moreover, idleness corrupts the soldier as much as the lack of payment sours him. I assure you that a little money and activity are the appropriate remedies for this evil." [6]

While desertion decimated the army, illness also took its toll. Carlier occupied his working hours with this care, bitterly commenting that the officers did nothing to assist their men. Fortunately, few of the sick died, although Carlier reminded the secretary of state that he had not been given funds for a hospital and was reduced to using the money set aside for unexpected expenses to provide medical services.[7]

Aside from these problems, the army in Luxembourg remained relatively peaceful. Carlier spent the month of July in the camp near Bitburg. The town had surrendered and the army demolished its fortifications, while the intendant busied himself with routine cares such as supervising a review of the troops. The number of soldiers was not exact, Carlier warned Louvois, for reviews were hard to make in an army camp, but the intendant assured him that it was an accurate enough basis upon which to pay the troops.[8]

5. Carlier to Louvois, 20 June 1667, A.G. A¹ 209, fol. 46.
6. Carlier to Louvois, 1 July 1667, A.G. A¹ 209, fol. 63.
7. Ibid.
8. Carlier to Louvois, 25 July 1667, A.G. A¹ 209, fol. 106; accompanying the letter is a copy of the review, "Éstat de la force des troupes."

Carlier's usefulness ended in August. On the first of the month, Créquy reported that the intendant had a violent fever, and the illness lasted the entire month. Therefore he played no active role in the latter half of the campaign. Yet his job remained the basic one already sketched in preceding chapters: he watched the army's finances, cared for its sick, paid the troops on the basis of exact reviews, oversaw the work on fortifications, and reported everything that happened in the army.[9]

Louis Robert played a similar role in Aumont's army. Like Carlier, he concentrated on the ordinary tasks of establishing a hospital, paying the troops, and supplying the army. His most pressing problem was the care of casualties. Although Aumont's forces were small in number, they participated in the capture of Bergues, Furnes, and Courtrai. Consequently the army had a large number of injuries. Even at the beginning of the campaign, Robert reported that there were so many wounded in the army that he could not keep track of the number; it increased every minute. In addition to the wounded, Robert's army, like other military forces in the seventeenth century, always had a number of sick whose illnesses were commonly diagnosed as "fever" or "bloody flux." [10]

A hospital was needed for the injured and incapacitated, but Robert did not have such an institution in the modern sense. He merely billeted the wounded at three locations in the city of Bergues: the convent of the Filles Hospitalières, an order of nursing sisters, and two meeting halls of the harquebus confraternities. The intendant asked the city to furnish the beds necessary for the sick, and he arranged for the sisters to care for all three sites at the rate of six *sols* per day per soldier, hoping to save expenses and increase efficiency at the same time. A nursing convent could not be expected to provide all medical services for the wounded. This chore fell to two surgeons, an apothecary, several barber-surgeons, and a doctor from the neighborhood who had sometimes treated Monsieur the Prince. Robert also obtained an almoner and various valets from the army to assist the hospital. But since an intendant's responsibility lay with the army, Robert could not remain at Bergues indefinitely, so he named one of his *commissaires des guerres*, Monsieur Aubert, to supervise the hospital in his absence.[11]

9. Créquy to Louvois, 1 August 1667, A.G. A¹ 209, fol. 112; Carlier to Louvois, 30 August 1667, A.G. A¹ 209, fol. 156.

10. Robert to Louvois, 7 June 1667, A.G. A¹ 209, fol. 30.

11. Robert to Louvois, 10 June 1667, A.G. A¹ 209, fol. 35. To the tourist's delight, these harquebus confraternities still exist today in cities like Bruges, Belgium. For a precise English-language clarification of the difference between

Hospital care was expensive. Robert estimated the cost at Bergues to be ten *sols* per day per soldier, for over two hundred men. This amounted to three to four thousand *livres* a month. He reckoned the wages of the doctors, surgeons, apothecary, almoner, valets, and other orderlies at fifteen hundred *livres* per month. And he did not forget the cost of medicine, the "vinegar, wine, and alcoholic spirits," to clean and disinfect wounds. The sum, Robert suggested, would be no less than six thousand *livres* per month, and more if the number of wounded increased. Even then, he did not include the various medications, the chapel, and the forty beds with linen that he had already purchased in Paris and Dunkirk. This had cost another three thousand *livres*. It is not surprising that the intendant pleaded in letter after letter for Louvois to send more money.[12]

No additional money arrived for the hospital, however. The intendant hoped that Le Boistel, the receiver of contributions in Picardy and Artois, would be able to raise ten thousand *livres* on credit, but he received news from Dunkirk that Le Boistel had been unable to do so. Robert wrote: "the hospital subsists only from the purse of Aubert, which will undoubtedly soon be exhausted. I sent messengers to him [Le Boistel?] yesterday and today, with notes and letters of all sorts to enable him to find credit, but I doubt that this will succeed; meanwhile, all the unfortunate perish. I very humbly beg you to think of the importance of sending funds immediately to M. Aubert." Evidence is lacking on whether Louvois sent the money, but on 7 July 1667 Robert once again reminded the secretary to send ten thousand *livres* to Aubert.[13]

The main hospital remained at Bergues, but as the army continued its march, the intendant found it necessary to make other arrangements for the care of the wounded. In July, Robert sent his casualties to Armentières, appointing a commissioner Eyraut to supervise their care. Transporting the injured was a slow task, for the army needed carts to convey some ninety-four wounded soldiers. The intendant sent Eyraut what money he could, but he protested to Louvois: "Although you complain that the expense of the hospital climbs too high, I do not believe that it is your intention to abandon the sick. I have used all the

doctors, surgeons, barber-surgeons, and simple barbers, see Leon Bernard's chapter, "The World of Medicine," in his *Emerging City: Paris in the Age of Louis XIV* (Durham, N.C., 1970), pp. 209–33, particularly pp. 217–18. "Estat des rations de pain pour les officiers d'artillerie de l'armée de M. le Mar.ᵃˡ Daumont," A.G. A¹ 209, fol. 42 (following a letter of Robert, dated 15 June 1667), mentions two *commissaires provinciaux*, four *commissaires ordinaires*, and six unspecified commissioners in the army.

12. Robert to Louvois, 10 June 1667, A.G. A¹ 209, fol. 35.
13. Robert to Louvois, last of June and 7 July 1667, A.G. A¹ 209, fols. 62, 76.

care and economy that I could. If you want it to cost less, only tell me, and I will regulate their good or poor treatment according to the expenditure you want to make." Later, after the fall of Courtrai, Robert left only the most seriously wounded in that town, since he did not have enough doctors—they were all at Bergues. He sent the rest of the wounded to Tournai. The lack of money, the continuous movement of the army, and the small number of casualties—only fifty to sixty were wounded at the siege of Courtai, but the army had another hundred who were sick—excluded the possibility of establishing more permanent hospitals. Instead, the intendant contented himself by bedding the sick in nearby towns.[14]

Lack of money remained a problem. Like Carlier, Robert impatiently awaited the arrival of money to pay the troops. The reason for the problem lay with Louvois's inexperience. Although Louis XIV's war was a military triumph, the young secretary had difficulty coordinating the demands for money and supplies. At first Robert was satisfied with a passing mention: "I do not speak at all of the payment of the troops, nor of the general staff [*état-major*], not doubting that you have seen to it and that the funds will soon arrive. . . . By not paying the soldiers, it is impossible to keep them in rigorous discipline." More insistent demands for money soon followed. The troops at Armentières had not been paid. Robert questioned what he should do every time the French captured a place and left a garrison there, such as at Armentières. Did the intendant have to pay them? If the reply was affirmative, the secretary should send the money immediately.[15]

The problem of money was closely connected to that of supplies. The documents do not present a clear picture of the army's victualing in this war, only hints of Robert's efforts. During the first week in June, the intendant requested that money be sent for the subsistence of the garrison lodged at Bergues; otherwise it would receive bread alone—a potential cause of disorder. There were problems even in obtaining bread for the troops. By 24 June, Robert announced from Armentières that he had received a shipment of sixty thousand rations of bread, fourteen hundred sacks of flour, and a hundred fifty *septiers* of grain to be milled. Most of the bread, he complained, was "crumpled, broken, and moldy [*presque tout froissé, rompu et moisy*]," amounting to a loss of more than thirty thousand rations.[16]

Even when flour arrived, there still remained the problem of transport. Louvois sent Robert carts from Turenne's army, but the in-

14. Robert to Louvois, 2 and 18 July 1667, A.G. A¹ 209, fols. 66, 100.
15. Robert to Louvois, 10 June and 7 July 1667, A.G. A¹ 209, fols. 35, 76.
16. Robert to Louvois, 7 and 24 June 1667, A.G. A¹ 209, fols. 29, 55.

tendant grumbled that the number was insufficient. Robert could transfer only twelve hundred *septiers* of grain to Tournai, whereas he had two thousand *septiers* in his possession. To obtain more carts, he sent orders into the surrounding countryside to requisition peasant wagons, but he could not estimate how much flour he could deliver to Tournai, although he doubted that it would be more than fifteen hundred *septiers*. Lacking carts, Robert could not ship the surplus to Arras, but he had given orders to send grain by canal barge as far as Saint-Venant.[17]

As Aumont's army advanced, Robert left Armentières, giving commissioner Eyraut funds to buy bread for the garrison remaining in that town. This was not enough, the intendant suggested to Louvois, and it was of the utmost importance to transport flour from Saint-Venant to Armentières. The government responded by shipping a thousand sacks of grain to the town, and ordering Aumont's army to obtain its bread via this supply route. This, Robert retorted, was absolutely impossible. The army was a good seven to eight hours from Armentières, therefore a two-day trip for a convoy to return with supplies. Moreover, the army caissons were only large enough to carry two days' rations. It would take a continuous stream of convoys back and forth to supply the army, and since the route passed near the enemy territory of Lille, Robert felt that convoys should not be exposed to this danger without a strong escort. Thus he suggested to Louvois that the secretary give him permission to take the bread from Tournai. Undoubtedly Louvois did, for Robert got his supply of bread from Tournai. Once the French had captured Douai, Robert could transport supplies from Armentières by the Lys River, so he sent some supply assistants and bakers to prepare bread at Armentières. This incident illustrates two things: the inexperience of the young Louvois, and the importance of having an agent in the field as opposed to a bureaucrat issuing orders from a desk.[18]

As an eyewitness, an intendant could supply immediate information to his master, the secretary of state for war. For example, in mid-July, Robert reported a dispute between the sieur de Peguilin, a brigadier, and the lieutenant generals of the army. Like many disputes, the disagreement arose over the question of precedence. Peguilin claimed the privilege of receiving the watchword for the day directly from Marshal Aumont instead of going through the lieutenant generals. He based his claim upon a letter from the king ordering him to command his troops as a separate corps, although under Aumont's authority. Robert

17. Robert to Louvois, last of June 1667, A.G. A¹ 209, fol. 62.
18. Robert to Louvois, 2, 4, and 7 July 1667, A.G. A¹ 209, fols. 66, 68, 76.

reported that the lieutenant generals had tried to work out a compromise: they would not give Peguilin the watchword, and he would not attempt to take it directly from Aumont; rather, he would receive it from the major of a brigade until the king had arbitrated the matter. Peguilin refused the accommodation. At this point Aumont intervened and ordered him to receive the watchword from the lieutenant general in command that day. The secretary's response is not known, but the point is clear: Robert, an army intendant, filled the traditional position of government informant, describing any event that might prejudice the army's progress.[19]

Secure in Louvois's confidence, Robert continued to act in an independent fashion toward the officers. He seemed to have forgotten his reprimand in the Coligny affair, and once again he drew upon himself the ire of a general. At the beginning of August, the intendant announced to Louvois: "I told Marshal Aumont that our idleness did not accord with the orders of the king." Aumont had cut him short, gruffly saying that he would not undertake any enterprise against the enemy until the return of a lieutenant general named Lorges. Robert had retorted that Lorges returned the evening before with new orders from the king to march. Embarrassed by the intendant's charge, the general replied that he would decamp the following day and prepare for action. But one of Aumont's friends privately told Robert that the marshal had no intention of moving from the location for the present time. This was only one instance, the intendant announced, of Aumont's anger, although Robert admitted: "He has not said anything to me [personally] but he strongly complained of me to several persons and he acts very cold toward me, not saying a word." This may be explained as a personality clash, but a more likely reason is that the intendants no longer remained as obsequious to a commanding officer. Confident of their position, they actively asserted their own opinions even if this meant incurring a general's rancor.[20]

Certainly Robert and Carlier performed the customary duties of an army intendant, but the War of Devolution saw a return to a previ-

19. Robert to Louvois, 11 July 1667, A.G. A¹ 209, fol. 87.

20. Robert to Louvois, 2 August 1667, A.G. A¹ 209, fol. 115. In an earlier letter to Louvois, 24–25 June 1667, A.G. A¹ 209, fol. 55, Robert also complained that Aumont did not follow his advice: "Je vous diray en deux mots que contre mon sentiment nous faisons icy bien plus de fortifficaõn que le Roy n'en a sans doubte intention, j'en aurois empesché beaucoup du dessein sans mon incommodité et je tasche tous les jours de le diminuer, mais il est apropos que vous recommandez encores un fois que ce ne soit qu'un camp retranché." It was quite a natural response for a general to be suspicious and taciturn toward an intendant who reported his every lapse to the secretary of war.

ous type of military intendancy. The army intendants gradually transformed themselves into intendants of newly conquered territories. This transition from a military to a quasi-military-civilian administration has no easy dividing line. Robert was an excellent example. While Aumont marched through western Flanders, the intendant gradually shifted his concern from the army to the problems of quartering soldiers upon a captive populace and surveying the newly conquered places.

For example, after Aumont's capture of Bergues in early June, Robert inspected the barracks already built by the Spanish government. "The caserns are very good [*fort belles*], entirely new [in fact]," he reported; "there are some already constructed for the infantry as well as the cavalry, but there are also some which are not yet finished, but well advanced." The woodwork was partly completed, and the structures needed roofs. The Spanish had had plans for enough barracks to hold three thousand men; the French had an occupation force of five hundred in the town. Robert wondered whether Louvois wanted to finish construction on these caserns. The intendant also had to supply the soldiers. He requested through Aumont that the civilians provide beds and two *sols* per day to each soldier. Because Robert feared the possibility of trouble arising from a poorly paid garrison in a newly conquered place, he ordered the military commissioner, Aubert, to conduct the requisition of beds and linen "in a gentle fashion," ceasing entirely if it agitated the inhabitants too much.[21]

Robert also inspected the town of Bergues for strategic possibilities. He found that there were several breaches in the city walls, but he felt no hurry in fixing them because the enemy was too weak to attempt a reconquest. Robert reported that a canal was under construction, extending from Bergues to the town of Furnes. Since the canal would enrich Bergues at the expense of Dunkirk's commerce, the intendant advised the destruction of the canal if the crown decided to return Bergues to Spain at the end of the war, but if the king decided to keep Bergues, he should preserve the canal or face the wrath of the inhabitants. For the present, Robert added, he had given his permission to continue work on the canal until he received further advice from the court.[22]

The city had asked in its capitulation that all its privileges granted by the Spanish monarchs be confirmed. The intendant agreed to the request, but added the provision that this would be only so long as the privileges were not contrary to the French king's prerogatives. He also adjusted one article of the capitulation, separating Vieux-Mardyck, Petite-Sainte, Grande-Sainte, and Zucotte with their de-

---

21. Robert to Louvois, 10 June 1667, A.G. A¹ 209, fol. 35.
22. Robert to Louvois, 7 and 10 June 1667, A.G. A¹ 209, fols. 29, 35.

pendencies from the castellany of Bergues and joining them to that of Dunkirk. Robert believed that this was best because it might be possible that Bergues would be returned to the Spanish without Dunkirk, but never Dunkirk without Bergues. Consequently it was better to keep the villages united to Dunkirk, where French control was more assured. The citizens of Bergues vigorously protested the intendant's action, but he quieted them with the promise that they could appeal his decision to the king.[23]

Along with Robert, a swarm of other royal officials descended upon the newly conquered territories. After September 1667, these men received commissions as "intendants of contributions" with the task of administering the lands that had fallen into French hands as a result of the summer campaign. One of the first to arrive—and surely the most remarkable—was Jacques Charuel. At the end of July he had entered the French zone of occupation and acted as Louvois's agent by conferring with local magistrates on the levy of forage and the site of future forage magazines, inspecting the conditions of the citadels of Courtrai and Armentières, and negotiating the amount of "taxes" the newly acquired towns would have to pay. At Audenarde, for example, Charuel reached an agreement with the local tax farmers by which they submitted to French authority in return for keeping their jobs. During these months it was impossible for Charuel to have been an intendant in this area, for the territory had not yet been divided into intendancies. Yet if he did not have the title de jure, Charuel certainly acted as de facto intendant.[24]

If one of his letters, dated 6 September 1667, is any indication, Charuel certainly played an important role in the delicate job of partitioning the conquered lands into intendancies. In the letter, he announced to Louvois that he could not send immediately the proposed project of intendancies that the two had agreed upon, because it remained with the rest of his papers at Tournai. But Charuel promised that he would send the proposal as soon as possible.[25]

23. Robert to Louvois, 10 June 1667, A.G. A¹ 209, fol. 35. With the letter, the intendant wrote that he was sending a memorandum of all the gifts, concessions, and privileges granted by the Spanish crown to the city in return for the construction of the canal and caserns, as well as a list of taxes and other revenue normally paid to His Catholic Majesty. This list, however, is missing from *recueil* A¹ 209. The marquis Marie de Roux, *Louis XIV et les provinces conquises* (Paris, 1938), p. 128, remarks that the capitulations signed by the captured towns repeat the same provisions and were often copied from each other.

24. See Charuel to Louvois, undated (probably last of July), 5, 25, and 30 August, and 3 September, A.G. A¹ 209, fols. 111, 119, 146, 154, 159.

25. Charuel wrote to Louvois, 6 September 1667, A.G. A¹ 209, fol. 168: "Je ne puis presentement vous envoyer le projet de separon des intendancies des levées de Flandres que j'ay fait avec vous, non plus le memoire juste des choses que j'ay fait jusques a present par ce que ces papiers sont demeurez a Tournay."

After the division had been made, Charuel received the largest intendancy. His jurisdiction extended over the castellanies of Lille, Tournai, and Courtrai, and included the land on the other side of the Escaut River belonging to Audenarde up to Ath and Alost. Charuel was not happy about its size, and he suggested that Louvois give the towns of Audenarde, Tournai, and Ath to Talon because he doubted whether he could supervise all these territories by himself. "Do me the justice," Charuel argued, "of believing that I am not backing out of work—the affairs of all the places are almost regulated and established —but I certainly see that I cannot act everywhere this winter, and they [the towns] absolutely must have the presence of a man armed with primary authority and the backing of the court." Louvois reluctantly agreed to part of Charuel's proposal, giving Audenarde to Talon, though he grumbled: "I wished very much that you could have undertaken the whole extent of the employ that the king has given you, for many reasons that you can judge as easily as I." [26]

Perhaps Louvois had doubts about Talon, for he was careful to make him dependent on Charuel's authority. It was Charuel who met Talon at Courtrai to discuss the subject of Talon's intendancy. Afterward, Charuel reported to Louvois that Talon had agreed to work together for the common good of the troops. A few days later, Louvois notified Charuel that he had sent Talon his commission as intendant of Audenarde, including all the territory between the Escaut and Lys rivers not belonging to Charuel's intendancy. The secretary noted that he had not sent Talon any instructions: "I only told him that there must be a perfect intelligence between you two, and that during the very long time when I had been with you during the [present] year, I had amply instructed you on His Majesty's intentions regarding the contributions. You will explain to him the limits of his department and what he is to do there." The next year, when there was a general reshuffling of intendancies after the Peace of Aix-la-Chapelle, Talon re-

26. Charuel to Louvois, 16 September 1667, A.G. A¹ 209, fol. 199, and Louvois's reply, 23 September 1667, A.G. A¹ 208, fol. 18. I did not find Charuel's commission in the war archives, but various letters, for example that of Louvois, 23 September 1667, A.G. A¹ 209, fol. 17, bear the notation "intendant en Flandres." It does not appear that Louvois accepted Charuel's suggestion in its entirety, for he granted only Audenarde to Talon and allotted Tournai, Lille, and Courtrai to Charuel. See the letter to Charuel, 23 September 1667, A.G. A¹ 208, fol. 17, in which Louvois defined the limits of Talon's intendancy: "j'envoye par cet ordinaire a M. Talon sa commission d'intendant a Oudenarde, a laqᴸᵉ je serois d'avis de joindre tout ce qui est entre Lescaut, et la Lis qui n'est point de la chastellenie de Lisle du Tournaisis, ny de la chastellenie de Courtray, et de l'autre costé dudit Escaut tout ce qui est de la chastellenie d'Oudenarde, et ce que vous n'avez point attaché a la recepte d'Ath qui n'est pas de la chastellenie d'Alost."

mained subordinate to the new intendant of Flanders, Michel Le Peletier de Souzy.[27]

There were other agents besides Charuel and Talon who became intendants in the occupied frontier lands. Terruel, seigneur d'Estrepigny, received a commission, dated 28 September 1667, as intendant of contributions in Le Quesnoy and the towns between the Sambre and Haine rivers up to Binche, and from Thuin to the Scell River. Jacques Camus des Touches, who had served as *commissaire des guerres* under Robert during the Hungarian expedition of 1664, became the intendant of contributions in Hainaut. His department included the towns of Charleroi, Philippeville, Mariembourg, Avesnes, and Charlemont. Louvois did not forget Robert and Carlier, either. Etienne Carlier accepted a commission as intendant of contributions in Luxembourg; his jurisdiction extended throughout the lands of Luxembourg and in all the districts situated along the Meuse River, except for the county of Namur and parts of Guelders and Limbourg that were under Camus des Touches's control. Louis Robert's commission assigned him the castellanies of Furnes, Bergues, and Cassel, along the seacoast.[28]

27. Charuel to Louvois, 16 September 1667, A.G. A¹ 209, fol. 199; Louvois to Charuel, 23 September 1667, A.G. A¹ 208, fol. 17. Talon's commission, undated but probably 23 September 1667, names him "pour avoir l'intendance et direction des levées que nous voullons et entendons estre doresnavant faites tant en argent, qu'en especes comme grains, foins, pailles, et autres denrées necessaires pour la subsistance des troupes de nos armées de Flandres, et sur les villes, bourgs, villages et lieux des chastellenies d'Oudenarde et Alost," A.G. A¹ 208, fol. 22. In May 1668 when Louvois announced the change of intendancies, he wrote to Charuel, 10 May 1668, A.G. A¹ 222, fol. 530: "La chllñie d'Oudenarde demeure a Mʳ Talon avec la mesme subordination a M. de Souzy." Concerning Louvois's later suspicions of Talon, see his letter to Charuel, 6 January 1668, A.G. A¹ 222, fol. 16. This member of the Talon family is difficult to identify; it is probably Claude Talon, for he is named in this manner in B.N. *collection de Chérin* 192, family "Talon," fols. 2–6.

28. Commission of Terruel, 28 September 1667, A.G. A¹ 208, fol. 31: "pour avoir l'intendance et direction des levées que nous voulons et entendons estre doresnavant faites tant en argent qu'en especes comme grains, foins, pailles, et autres denrées necessaires pour la subsistance des troupes de nos armées de Flandres, et ce sur le Quesnoy et sur les villes, bourgs, villages et hameaux, et aües lieux sçituez entre les rivieres de Sambre et d'Haisne a prendre depuis Binche, et Thuin jusques a la riviere de Scelle." Terruel also entitled himself "le Sʳ de Terruel, Chlĕr seigneur d'Estrepigny conᵉʳ du Roy et intendant des levées en deniers ou en especes pour la subᶜᵉ des troupes, et l'entretennement des places frontieres dans une partie du Haynaut et du Cambresis," in an order during October 1667, A.G. A¹ 209, fol. 301. Information on Terruel is very difficult to obtain. I found no trace of him in the genealogical records in the Bibliothèque Nationale, but he is probably the same Terruel who served on the expedition to Hungary in 1664.

There is no copy of Camus des Touches's commission in the war archives, but a letter of Louvois to Camus des Touches, 27 September 1667, A.G. A¹ 208, fol. 28, mentions that he is to obtain contributions from Charleroi, Avesnes,

Men such as Robert, Carlier, Camus des Touches, and Charuel were professional administrators. They had served the war department for most of their careers. Yet in this world of patrons and clients, court influence could secure relatively unknown men an intendancy in the newly conquered area, showing that expertise was not always a necessary requirement. In October 1667, Louvois informed Charuel that "Monsieur Gaboury, who has served the late queen mother for a long time, desired to have an employ so much that I could not resist according him that of the *bailliages* of Aire and Saint-Omer with Saint-Venant, the latter to serve as his residence." Gaboury received his commission as intendant of contributions on 21 November 1667, the last of the intendancies granted after the first year's compaign.[29]

Philippeville, Mariembourg, and Charlemont. Various letter headings in *recueil* A¹ 208, such as that of Louvois, 27 September 1667, A.G. A¹ 208, fol. 28, refer to him as "intendant en haynaut." For the extent of his intendancy, see Louvois's letter of 10 December 1667, A.G. A¹ 208, fol. 257.

Carlier's commission, 16 October 1667, A.G. A¹ 208, fol. 97, names him "pour avoir l'intendance et direction des levées. . . . dudit pays de Luxembourg et dans tous les lieux qui sont sçituez audela de la Meuze apartenant au Roy catholique a la reserve toute fois de tout ce qui depend du Comté de Namur, et de quelque partie de la Gueldre et du Limbourg qui contribuent a Charleroy."

Robert's commission, undated but probably 8 October 1667, A.G. A¹ 208, fol. 63, names him "pour avoir l'intendance et direction des levées . . . sur les villes, bourgs, villages, et autre lieux depend^t des chastellenies de Furnes Bergues et Cassel."

The information on Colbert de Croissy is obscure. He was the only person who served in the campaign who did not get a military intendancy. He remained with Turenne's army until its return in late October. Then he traveled to Amiens to supervise the supplying of the troops that passed through on their way to winter quarters. As Colbert's brother, moreover, he was in a different class from the rest of Louvois's protégés. At the end of 1667 it seems that he became intendant of Paris, for both Pierre Clément, ed., *Lettres, instructions et mémoires de Colbert* (Paris, 1863), 2, pt. 1: CCLXXIV–CCLXXIX, and Charles Godard, *Les Pouvoirs des intendants sous Louis XIV particulièrement dans les pays d'élections de 1661 à 1715* (Paris, 1901), pp. 518–40, mention him in their lists of intendants of Paris during this period. But Jean Bérenger in his article on Croissy, in Mousnier, *Le Conseil du roi*, p. 159, states that he was intendant of Flanders in 1668. In any case, the end of the War of Devolution saw his transfer to London as French ambassador, where he remained until 1674.

29. Louvois to Charuel, 14 October 1667, A.G. A¹ 208, fol. 81. Gaboury's commission, 21 November 1667, A.G. A¹ 208, fol. 297, allotted him jurisdiction "dans les villes bourgs et villages et lieux deppend^t des bailliages de St Venant, Aire et St Omer regalle de Terroüanne et lieux de ce costé la ou vous pourréz estendre lesd. levées." Probably because of Gaboury's inexperience, Louvois did not name him intendant right away, for the secretary's above letter to Charuel continued: "je vous prie de m'envoyer un memoire qui puisse luy servir d'instruction; et non seullement jusques là, mais encores jusqu'a ce que vous [Charuel] avez reglé avec Mons. le duc de Bournonville le traitté des centiesme d'Arthois, je ne luy deslivreray point sa commission, ny ie ne le feray point partir." In any event, the intendancy was of short duration, for the Peace of Aix-la-Chapelle in spring 1668 handed this area back to Spain.

The boundaries of these intendancies were vague. Often they were enclaves inside Spanish territory. Consequently, their commissions assigned the intendants towns and their dependencies without clearly stating where their jurisdiction stopped and another's began. Conflict often broke out among the intendants because each felt another was interfering in his department. Only Louvois could settle these quarrels. At the end of September 1667, the secretary advised Camus des Touches to be careful not to surpass the limits of his intendancy, which was along the frontier of Cambrésis, for that area should be left to Charuel. If Camus des Touches had intervened in the district, any changes he made should be rescinded to avoid confusion. Moreover, Louvois warned him, he should not levy contributions in Le Quesnoy, Terruel's intendancy, "which, being of very tiny extent, would become nothing if it was diminished." [30]

Camus des Touches was not the only intendant to interfere in his neighbor's department. When Charuel complained that Talon had overstepped his bounds, the secretary promised, "I will write to M. Talon so that he will no longer interfere in what regards the *vieux bourg* of Ghent or the *franc* of Bruges, without revealing that you have written me." A much more serious difference broke out between Carlier, intendant of contributions in Luxembourg, and Jean-Paul de Choisy, provincial intendant of the three bishoprics of Metz, Toul, and Verdun. In mid-November, Louvois tried to arbitrate the differences between the two. He told Carlier: "M. de Choisy can and should levy the imposition in all the lands in the expanse of the generality of Metz and in those that were under the obedience of the king before the month of last May; and, you, you must levy impositions only upon the places subordinate to the enemies in those times. Regarding the furnishing of forage to the troops, you must deal fairly [*fort honnestement*] with M. de Choisy. . . . Have no contestations with him, no matter what happens. If it is necessary to satisfy him, take the rank of subdelegate." Evidently this warning did not suffice, for Louvois wrote again in December. This time the tone was much sharper: "I

30. Louvois to Camus des Touches, 27 September 1667, A.G. A¹ 208, fol. 28. The best maps of the area are found in Nelly Girard d'Albissin, *Genèse de la frontière franco-belge: Les Variations des limites septentrionales de la France de 1659 à 1789*, Bibliothèque de la Société d'histoire du droit des pays flamands, picards et wallons, no. 26 (Paris, 1970), at end of volume. The maps show the incredible political complexity of the region with its overlapping boundaries, disputed territory, and jointly occupied areas (*terres mi-parties*). Although Girard d'Albissin's interest is not, strictly speaking, with the details of the occupation, her work is crammed with details about French policy and its results on the frontier. See particularly chap. 2, "La Situation militaire et économique en Flandre et en Hainaut, après la paix d'Aix-la-Chapelle," pp. 101–57.

have already told you that there must be no contestation with M. de Choisy, so do not take over the care of the confiscated goods in the expanse of the territories that were obedient to the king during peace-time except as his subdelegate." [31]

The boundaries of these intendancies, of course, were not meant to be permanent. They had been erected at the close of the first year's campaign, and Louvois did not hesitate to transfer large blocks of territory as the military situation changed. In December 1667, the king decided to grant Créquy's request that he, instead of M. de Montal, should have control over the former Spanish territory along the Meuse River. This meant that Carlier, intendant of the conquered area where Créquy was in command, would have his jurisdiction extended to these areas. And part of the county of Namur between the Sambre and Meuse rivers was transferred to the jurisdiction of Philippeville, which was in Camus des Touches's department. These transfers were only temporary; the conclusion of the peace the following summer resulted once more in a reshuffling of territories and the reassigning of intendancies. [32]

The 1668 campaign began in February with the whirlwind conquest of the Franche Comté. Despite the diplomatic pressure from the newly formed Triple Alliance of England, Sweden, and the Netherlands to end the war and prevent further French acquisitions, Louis XIV decided to put three more armies into the field later that spring: one under the command of the king in person, assisted by Turenne; a second under Monsieur, assisted by Créquy; and a third under Condé to operate in Germany. The crown named two new men as army intendants. Louis-François Le Fèvre de Caumartin, provincial intendant of Champagne, obtained the intendancy in the army of Germany, while Paul Barillon d'Amoncourt, a former intendant of Paris, accepted a commission in Turenne's army in Flanders. Louis Robert received orders to serve in Créquy's army, and he left his department of Furnes, Bergues, and Cassel along the seacoast to the care of his *commissaire des guerres*, Aubert. Charuel, Talon, Terruel, and Camus des Touches remained intendants of contributions in the occupied territories of Flanders, probably because the government did not want a radical change in administration before the campaign. [33]

31. Louvois to Charuel, 25 November 1667, A.G. A¹ 208, fol. 213; to Carlier, 16 November and 9 December 1667, A.G. A¹ 208, fols. 178, 249.

32. Louvois to Camus des Touches, 10 December 1667 and 5 January 1668, A.G. A¹ 208, fol. 257, and A¹ 222, fol. 9; to Carlier, 5 January 1668, A.G. A¹ 222, fol. 8.

33. Rousset, *Histoire de Louvois*, 1:128–29, 140. I found no evidence of any army intendant in the expedition to conquer the French Comté in February 1668,

The army's hope for a new campaign, however, was dashed by the actions of the diplomats. Already in March a truce had suspended hostilities for the duration of that month, and the negotiators pursued this advantage, despite setbacks, until they concluded the armistice of Saint-Germain by mid-April 1668. This agreement brought France to the peace table. About two weeks later, the French representatives signed the Treaty of Aix-la-Chapelle (2 May 1668), definitely ending the war. By this treaty Louis XIV gave up his winter conquest of Franche Comté, but he retained the occupied towns of Flanders: Lille, Bergues, Furnes, Armentières, Courtrai, Menin, Douai, Tournai, Audenarde, Ath, Binche, and Charleroi. The government ordered the demobilization of the army into several smaller corps to observe the fulfillment of the treaty, and toward the end of May, Robert and the two other army intendants returned to their peacetime duties.[34]

Once again the secretary of war shifted the intendancies within the occupied zone of Flanders, this time to conform to the needs of what he hoped was to be a permanent French annexation. The government constructed a larger intendancy of Flanders from the smaller elements and committed its administration to Michel Le Peletier de Souzy, who had been recalled from his intendancy in the Franche Comté. He exercised the function of intendant of justice throughout Flanders, but in order not to dispossess the other intendants entirely, the crown assigned them the responsibility for policing the troops. With the control of the soldiers went the local administration of certain towns. Le Peletier de Souzy acquired control of the two most important towns of Flanders, Lille and Tournai, but Talon remained in Audenarde and Charuel kept the towns of Douai, Ath, and Courtrai. Camus des Touches saw his department reduced to Charleroi, and Terruel retained Le Quesnoy. Louvois even trimmed Robert's department,

but after the conquest, Le Peletier de Souzy served as intendant until the province's return to Spain that spring; see his commission awarded on 24 March 1668 and antedated to the end of the conquest, 19 February 1668, A.G. A¹ 222, fol. 385. I found no commission for Caumartin, Barillon, or Robert, but there are scattered references to them in the war archives. There is a letter of Louvois to Caumartin, 1 February 1668, A.G. A¹ 222, fol. 189, announcing his intendancy; see also Louvois to Carlier, 6 April 1668, A.G. A¹ 222, fol. 440, which mentions "Mons. de Caumartin qui est intendant de l'armée." Louvois to Robert, 7 March 1668, A.G. A¹ 222, fol. 293, gives him permission to leave his department to join the army. "Memoire pour servir d'instruction au Sʳ Duc de Rocquelaure," 20 April 1668, A.G. A¹ 222, fol. 483, speaks of Barillon as "intendant de la justice police et finances en l'armée de Sa Maᵗᵉ."

34. Gaston Zeller, *Les Temps modernes: De Louis XIV à 1789*, Histoire des relations internationales, under the direction of Pierre Renouvin, vol. 3 (Paris, 1955), pp. 29–30. For the background of the peace negotiations, see Rousset, *Histoire de Louvois*, 1:139–54. See also Louvois to Barillon, 1 June 1668, A.G. A¹ 222, fol. 685, announcing the disbanding of the troops.

limiting it to Bergues and Furnes. When Robert complained that his intendancy was a "very tiny place" without an important city, he received the right to act as subdelegate in the towns of Dunkirk, Gravelines, and Bourbourg under the authority of Barillon, now intendant of Picardy. Although there continued to be small modifications, these intendancies generally remained until the outbreak of the Dutch War in 1672.[35]

The duties of these intendants were varied. According to their commissions, they had the official title of *intendant des levées*—literally "intendant of collections"—although they were called in common parlance "intendants of contributions." The commissions authorized them "to have the intendance and direction of the collections that we desire and intend to be henceforth made, both in money and in kind, in grain, hay, straw, and other provisions necessary for the subsistence of the troops of our armies of Flanders." A commission then set forth the general principles that should govern their conduct: the levies should extend as far as they could reach, they should not be restricted to the areas actually submitting to French control, and the intendant should make a careful inspection of his department to see exactly how much tax the area could bear "with justice and equity."

A commission also mentioned certain specific duties of an intendant. He had to sign the rolls for the collections, making sure that each document specified how much each area had to pay and the terms; he had to insure that the inhabitants punctually paid their "contributions" in conformity with the French demands, obtaining assistance from the governors of nearby towns when necessary; he had to appoint trust-

35. Concerning Le Peletier de Souzy's appointment as intendant of the "pays conquis" in Flanders, see Louvois to Charuel, 10 May 1668, A.G. A¹ 222, fol. 530. Le Peletier was the rational choice for this intendancy, since he was a master of requests and had the legal background necessary as a provincial intendant. Louis Robert, however, also requested that his commission be that of an "intendant of justice," because he was a "gradué"; see his letter to Louvois, 16 May 1668, A.G. A¹ 226, fol. 111.

Charuel accepted the reduction of his intendancy with good grace, telling Louvois: "je serviray le Roy et vous obeiray dans les postes que vous me destinez avec toute la fidelité, la passion, et le devoüement que je dois," 15 May 1668, A.G. A¹ 226, fol. 108, while Le Peletier, as might be expected, was effusive in his thanks; Le Peletier to Louvois, 10 May 1668, A.G. A¹ 226, fol. 62.

Regarding the other intendancies, see Louvois to Talon, 7 May 1668, A.G. A¹ 214, fol. 17; to Camus des Touches, 22 May 1668, A.G. A¹ 222, fol. 574; Robert to Louvois, 16 May 1668, A.G. A¹ 226, fol. 111, and Louvois's reply, 22 May 1668, A.G. A¹ 222, fol. 572. The crown replaced Camus des Touches with Etienne Carlier as intendant of finance and fortifications at Charleroi and Binche. Later in the summer the government amplified Carlier's commission to include other places in Hainaut, such as Avesnes, Philippeville, Le Quesnoy, and Mariembourg, which had formerly belonged to Camus des Touches's and Terruel's intendancies. Louvois to Carlier, 21 June 1668, A.G. A¹ 223, no. 129; to Terruel, 31 July 1668, A.G. A¹ 223, no. 292.

worthy people to receive the payment of money and provisions and observe their conduct in order to prevent any disorder in the collections; he had to establish magazines to store the accumulated supplies and to supervise their transport and distribution to the troops. This too entailed the appointment of honest guards. But the secretary of war gave the intendants considerable latitude in carrying out these orders, for even after the taxes had been fixed, the intendant could "reduce or increase them according to what you see, in conscience, needs to be done," and in general he could do "all that you judge appropriate for the greatest good of our service." [36]

Several copies of the instructions given to the intendants to supplement their commissions are extant. These documents are important because they clearly state the French objectives in the occupied area. Terruel's instructions were most explicit: "Tax each district according to what it can justly furnish His Majesty, whether in money or in hay, straw, and grain, observing not to tax them too much, but as much as they can bear so that they will be unable to pay the Catholic King [Charles II] any of the impositions that he levied on them before the declaration of war." Louvois earlier told Charuel the same thing in more colorful terms: "That is to say, one must raise as much money as possible, particularly in the castellany of Alost, which the Spanish will devour if we do not."

The demand to get as much money as possible out of the conquered territory was tempered only by the admonition to learn the relative strengths of each area in order to apportion the demands to the inhabitants' ability to pay. Despite this attempt at equity, Louvois saw no point in scaling down the French demands to the tax rates of the previous masters, the Spanish. The French, he wrote, could not gain the affection of the people of Flanders by taxing them lightly. Therefore one was justified in getting as much money as one could. There would be no exceptions to the French levies; Louvois specifically instructed Terruel that the rich abbeys and other ecclesiastical property be taxed the same as the lay territory.[37]

36. These powers are all stipulated in the commission of Robert, A.G. A¹ 208, fol. 63. The commissions of Talon, Terruel, and Carlier are almost exact copies except for slight changes in the preface. The phrases "intendant des levées" and "intendant des contributions" were used indiscriminately in the correspondence. For an example of the latter usage, see Louvois to Camus des Touches, 27 September 1667, A.G. A¹ 208, fol. 28: "je regleray au premier jour l'estendüe de l'employ d'intendant des contributions du Luxembourg que j'ay destiné a M. Carlier." Girard d'Albissin, *Genèse de la frontière franco-belge*, p. 108, confirms that the French were no respecters of the Spanish border: "Quant aux passages en force et à main armée à travers les terres d'Espagne, on les multipliait alors sans la moindre vergogne."

37. There are copies of only two instructions in the war archives for 1667: "Instruction au Sʳ Terruel pour faire les fonctions d'intendant des levées," un-

The French had two reasons for demanding as much money as possible. First, the government needed every bit of cash it could find to finance its war machine. When the king wrote to the governors of the occupied areas announcing the arrival of the intendants, he stated that their job was to raise the money "to use for the subsistence of my troops which I am obliged to station there in winter quarters." The idea that conquered lands should have to pay the cost of their own occupation was not a new one, although Louis XIV's government would push it in later years to the point of extortion.[38]

The second reason for the French impositions was that Louvois knew as well as Colbert that one could wage war by means other than armies. From the abstraction that any money obtained from the inhabitants prevented it from reaching Spanish hands, it was only a short step to this imperative: bleed the conquered lands dry so they cannot assist the enemy. "One must get all that one can," Louvois warned Camus des Touches at the beginning of his intendancy, "because when one does differently it only serves to give them [the conquered lands] the means to satisfy the demands that the Spanish will make upon them." Louvois's mind seemed fixed upon the idea that if the French did not get it, the Spanish would. This was not a hasty conclusion, seeing that the two countries were still at war until the spring of 1668.[39]

Since the impositions were important to the conduct of the war, Louvois watched these levies carefully. When the secretary noticed that Camus des Touches had obtained only twenty-five thousand *écus* for a six-month period from the lands around Charleroi, he registered his dissatisfaction that a place of this importance produced so little, and admonished the intendant to reflect upon this consideration. The secretary expressed similar discontent with Carlier, declaring the

dated but probably 27 or 28 September 1667, A.G. A¹ 208, fol. 32; and "Instructions a M Gaboury allant a St Venant pour les contributions dAire et de St Omer," undated but probably after 16 December 1667, A.G. A¹ 208, fol. 298. Louvois to Charuel, 23 September 1667, A.G. A¹ 208, fol. 17.

38. The king to Nancré, governor of Le Quesnoy, 28 September 1667, A.G. A¹ 208, fol. 33. See a similar letter of the king to Grancé, governor of Thionville, 18 October 1667, A.G. A¹ 208, fol. 96. The notorious example of French extortion was their demands upon the occupied territories in Holland during the Dutch War; see Rousset, *Histoire de Louvois*, 1:435–40.

39. Louvois to Camus des Touches, 27 September 1667, A.G. A¹ 208, fol. 28. See also the similar letter to Carlier, 16 November 1667, A.G. A¹ 208, fol. 178, in which Louvois remarked: "que si vous demandiez la contribution un peu forte par des envoys par⸢ers⸣ vous mettriez la pluspart des lieux hors d'estat de rien payer au Roy d'Espagne au cas qu'ils satisfeissent a vos envoys, et en cas qu'en n'y satisfaisant pas ils desertassent vous priveriez encores les Espagnols de leurs secours, et ils ont beaucoup moins de moyens de s'en passer que le Roy."

proposition that Luxembourg pay only fifty thousand *écus* for one year's contribution was "absolutely out of reason." But he admitted that he had left it to the intendant and the governor, the marquis de Créquy, to get as much money as they thought possible. If they believed they could not get more than a hundred thousand *écus*, Louvois would support them, but they were still to try their best. The secretary even offered the gratuitous suggestion of double taxation: levying the French imposition and then taxing the sum that the Spanish normally collected. Nevertheless, Louvois urged the intendant and the commanding general of the occupation forces to do what they thought most appropriate in the conquered territories.[40]

The letters of Louis Robert, intendant in the castellanies of Furnes, Bergues, and Cassel during the winter months of 1667–68, give some indication of the method of collecting contributions. Early in October he made a personal inspection of the lands under his jurisdiction. With the knowledge gained, he began bargaining with the local authorities in each of the three castellanies to fix the amount of money and supplies that they would agree to furnish the French king. On 12 October 1667, Robert reported to Louvois that he had done everything possible to get the magistrates of Bergues to agree to the contribution. He had employed all sorts of blandishments, even promising to use his influence to see that fewer than four regiments of cavalry were quartered in the town; otherwise, the intendant threatened, unless Bergues cooperated, a large number of cavalry would be stationed in the town.

Bergues protested that its revenues had decreased because the French had taken away their territories of Petite-Sainte, Grande-Sainte, Zucotte, and Vieux-Mardyck and joined them to Dunkirk. Already, the city fathers complained, the town had spent some fifty thousand *livres* for fortifications, reparations, and forages. The stables needed to house the French cavalry would cost much more, as would the bedding for the infantry and cavalry. The hay crop had been small that year due to drought. If the area had to use what little forage it possessed to feed the king's horses, their own cattle would die of hunger. Furthermore, the local economy was shaky; the export of butter, one of the area's main products, suffered from a drop in prices caused by the English flooding the market with cheaper butter.

40. Louvois to Camus des Touches, 27 September 1667, A.G. A1 208, fol. 28; to Carlier, 16 November 1667, A.G. A1 208, fol. 178. Although the intendants negotiated the exact sum of the contribution with representatives from the conquered areas, the actual collection of the imposition was left in the hands of specially appointed tax collectors called "receivers"; but the intendant had to file periodic reports on the receipts with Louvois.

After listening to these complaints, Robert tried to bargain once again. Since he had deliberately inflated the French demand for contributions, now he offered to reduce the amount to a more reasonable sum, the sum he had hoped to obtain all along—two hundred forty-five thousand *livres* in money and supplies. The magistrates refused even to contemplate this figure, pretending that it was absolutely impossible for them to offer even sixty thousand *livres*. This figure, in turn, was completely ridiculous to Robert. The town fathers grudgingly conceded, but they were not yet ready to set a final figure.

Since the area had to pay both money and supplies, Robert tried a different tack. According to Louvois's instructions, the army needed 242,300 rations of forage; so, unknown to the magistrates, the intendant rounded this figure off to a flat demand for two hundred fifty thousand rations. Again the magistrates proved obdurate. They argued that it was impossible to furnish this amount, although Robert forced them to admit that the land had not been ruined by the passage of French troops. The intendant replied that there was enough hay for the cavalry horses in the countryside, and the cattle could eat the large quantity of straw (*paille*) resulting from the recent harvest. When the officials protested that there was little grain in the district to feed the horses, Robert suggested that it would be easy to import from neighboring lands where the price was cheaper. Valuing forage at ten *sols* per ration, Robert estimated that the total would be a hundred twenty-five thousand *livres*. If this sum were deducted from the intendant's demand for two hundred forty-five thousand *livres*, the district would have to pay only a hundred twenty thousand *livres* in cash. Robert promised Louvois that he would remain firm in his demand for the total sum of two hundred forty-five thousand *livres*, although he admitted to the secretary: "To tell the truth, I do not think that they are in condition to furnish it." But he would not think of reducing the sum until he received Louvois's instructions.[41]

Bergues comprised only part of Robert's intendancy; he made similar demands for contributions from Furnes and Cassel. At Furnes he put forth a preliminary request for three hundred fifty thousand *livres* including forages. Although he had not received an official reply by 12 October 1667, the intendant reported that the townspeople said it was impossible to fulfill the French demands, for the Spanish governor of Nieuport had inundated their lands, drowning the forage in the countryside. Robert promised Louvois that he would investigate these complaints in person. The intendant made a similar demand for forage

41. This and the preceding paragraphs are taken from Robert to Louvois, 12 October 1667, A.G. A¹ 209, fol. 276.

from Cassel; the *bailliage* had to transport a hundred thousand rations of forage to the garrison at Armentières, a ration to be composed of fifteen *livres* of hay, fifteen *livres* of straw measured in *poids de marc*, and three *picotins* of oats, measure of Arras. In addition, Cassel had to furnish a hundred thousand bales (*bottes*) of straw weighing fifteen *livres* each.[42]

Although the castellanies grudgingly granted the provisions, they were parsimonious with their money. Robert had to confess to Louvois that he was behind in obtaining cash contributions. In all three places—Bergues, Furnes, and Cassel—he found the people deaf to his demands. He reported that they merely shrugged their shoulders and offered nothing, muttering that they saw that the French wanted to ruin them because the sums were unattainable no matter what violence they suffered. "I do not see," Robert wrote, "how to get one *sol* willingly," and he admitted that he did not know how to get the hundred thousand *écus* out of the occupied territories that intendant Charuel believed possible, or for that matter even two hundred thousand *francs*.[43]

Louvois was not very sympathetic to the castellanies' cries: "I have difficulty understanding how one can complain and remonstrate about the demands for money that you have made in the area." He knew that these people had been accustomed to paying considerable amounts in the past, and he wondered if there was someone fomenting rebellion. But Louvois accepted the intendant's suggestion that placing four regiments in Bergues and three in Furnes was too heavy a burden, and he agreed to withdraw one regiment from each castellany. At the same time, Robert received Louvois's instructions to ask for forage for the magazines: thirty thousand *livres* from Cassel, thirty thousand from Bergues, and fifty thousand from Furnes, payable by the tenth of November.[44]

Louvois authorized Robert to punish any village that refused to contribute. He suggested that the most profitable method of collecting the money was by seizing the livestock and selling it at the marketplaces of Furnes and Bergues. Profits from the sale would be handed over to the receiver of contributions. But Robert's inventive mind discovered a more efficient means of collecting forage. The provisions, he explained to Louvois, were not levied by his orders but by those of the local magistrates, for he had boldly faced the officials avowing that the

42. Robert to Louvois, 12 and 15 October 1667, A.G. A¹ 209, fols. 276, 300. The *livre* was the term for both a French money of account and a unit of weight.

43. Robert to Louvois, 15 October 1667, A.G. A¹ 209, fol. 300.

44. Louvois to Robert, 16 November and 22 October 1667, A.G. A¹ 208, fols. 176, 112.

French had no desire to commit any abuse; they only wanted supplies to feed their animals. To prove this to the officials, the intendant showed them copies of the reviews that listed the number of troops and how many supplies the horses consumed. By demonstrating the sincerity of his efforts and the necessity of the forage., Robert persuaded the local authorities that it was in their best interest to collect the provisions themselves. Getting the magistrates to cooperate with the army was the first step toward effective occupation.[45]

Other intendants of conquered territories faced problems similar to Robert's when they tried to collect forced contributions from the populace. As a result, quite often they were sympathetic to the area's complaints. Terruel, intendant of contributions at Le Quesnoy, wrote to Louvois in mid-November 1667 championing the poverty of his region: only thirty-eight villages in the district owed allegiance to his city; the rest belonged to Landrecies, which was part of Camus des Touches's department. The people had to pay to the French the heavy demands made upon them by their former masters: the *deux-vingtièmes* tax on their lands, hearths, and chimneys; an imposition on their horses and horned livestock; a tax on the consumption of meat and another on the brewing of beverages. Terruel added that these taxes were indeed a burden on people who had only their beasts and their labor for personal income. It would be difficult, the intendant insisted, to make them furnish the forage and grain needed for the regiment of Catheux during winter quarters.[46]

Louvois realized the problems in obtaining large sums from the conquered areas, but the government needed the money and could not be overly scrupulous. When the war department named Gaboury an intendant of contributions at Saint-Venant, it was quite specific about the measures to be used by the inexperienced intendant. Louvois carefully reviewed the taxes normally due from the two *bailliages* of Aire and Saint-Omer—the annual "two hundredth" tax, the *deux-centièmes* —but due to the intercession of the governor of Saint-Omer, the crown agreed to reduce this levy by half. In addition to this tax, Aire had to contribute thirteen thousand rations of forage to the magazine at Armentières. Collection of the funds was in the hands of Pingus, the receiver of contributions, but Gaboury had to supervise and provide pressure where needed, especially with regard to warning the inhabitants that the French would use martial law against those who refused to pay. To reduce the expense of dealing with recalcitrant taxpayers,

45. Louvois to Robert, 27 November 1667, A.G. A¹ 208, fol. 220*bis;* Robert to Louvois, 11 January 1668, A.G. A¹ 224, no. 73.
46. Terruel to Louvois, 24 November 1667, A.G. A¹ 210, fol. 31.

Louvois ordered Gaboury to make an example of one or two villages at a time. Even then the French were to follow strictly legal means; force should be used only after the intendant's approval of the receiver's request. The troops would then capture the village's livestock and sell it, the sum received being used to acquit an area of its obligation. Only in the last resort, when a village absolutely refused to pay, should it be treated as an enemy and given to the soldiers to sack.[47]

Obviously such acts required that the secretary employ elaborate safeguards to insure that all the moneys collected reached the king and were not dissipated among corrupt army officers and local officials. Louvois insisted repeatedly that it was the intendant, and only the intendant, who was to direct the impositions. "The king's intention," Louvois wrote to Gaboury, "is that only he [the intendant] can make demands upon the country, throughout the extent of his intendancy . . . and the governors have received orders to abstain from this." The secretary warned Charuel specifically: "in all the impositions that are levied for His [Majesty's] profit, carefully watch the conduct of the [army] commandants, and as the man who commands in chief has the reputation for being a model of regularity, see that those who serve under him follow his example." [48]

Such admonition was not an empty warning, for officers sometimes plundered villages for their own profit. A letter from Charuel dated 12 September 1667 reported the news that a lieutenant, Ducauroy, had pillaged the village of Noorkerke, despite the government's promise of royal protection given in return for its contributions. Charuel told the secretary that he had written to Ducauroy to make restitution for the sack. The intendant believed that it had taken place because of the example of a similar pillage by the governor of Saint-Omer after the nonpayment of contributions demanded by the governor.[49]

The collection of money and supplies formed only one of the economic weapons in the French arensal. The government also set about confiscating the goods of Spanish subjects, profiting from the sale of passports, and plundering occupied zones for raw resources, such as wood for the construction of gun carriages. Louvois repeatedly stressed the principle that the king of France must enjoy every right that the Spanish king had previously exercised in the area. The French carried this policy one step further: because the Spanish had ordered the confiscation of all goods belonging to the subjects of the French

47. Gaboury's instructions, A.G. A¹ 208, fol. 298.
48. Ibid.; Louvois to Charuel, 23 September 1667, A.G. A¹ 208, fol. 17.
49. Charuel to Louvois, 12 September 1667, A.G. A¹ 209, fol. 187. This was not the only complaint about army officers interfering with the contributions; see, for example, Camus des Touches to Louvois, 16 October 1667, A.G. A¹ 209, fol. 303.

king at the outbreak of the war, Louis XIV did likewise. All property belonging to avowed subjects of His Catholic Majesty was sequestered on the grounds that they were enemy aliens. The government placed this task in the hands of the intendants, who had to draw up lists of the quantity of goods owned and their estimated value; then they taxed the owners this equivalent sum. It is obvious that the French hoped to strike another blow against the Spanish, for Louvois's instructions stipulated that the intendants should force subjects near the border to pay "with the utmost rigor."[50]

Because the territory under French occupation was traditionally part of the Spanish Netherlands, its economic life was intricately bound to the area still in enemy hands. The French zone was more dependent upon the Spanish territory than vice versa, for its cities had traditionally exported their goods through ports that still remained in Spanish hands, while the important centers of economic life in the Spanish Netherlands were far from the occupied zone. Each side wanted to regulate the trade in this area to its own advantage through the use of customs duties and the manipulation of passports and safe-conducts. After a brief period of hesitation, the French decided to prevent commerce between enemy territories whose trade routes crossed through French territory. For instance, to prevent a merchant of Ghent in the Spanish zone from passing through French territory on his way to Bruges, another Spanish town, the French government decided to grant passports valid only for a certain route to a French occupied town. If the merchant traveled another route, his passport was invalid. In addition, if an enemy merchant wanted to visit more than one town in the occupied zone, he would have to purchase an additional passport for each town. In order to maintain tighter control over the traffic of goods, the French government ordered the river Lys closed to commerce.[51]

This elaborate system of passports was carefully regulated by the crown. Louvois issued instructions to the intendants governing the procedure: the lieutenant generals in the occupied zones had the authority to issue passports in the king's name, countersigned by the intendant. Of course, to allow the generals the right of issuing passports meant that they might manipulate the prices for their own profit, holding back considerable sums originally destined for the king. Without letting it appear that they were involved, Louvois suggested, the

50. Louvois to Charuel, 23 September 1667, A.G. A¹ 208, fol. 17; to Carlier, 18 October 1667, A.G. A¹ 208, fol. 95; Terruel's instructions, A.G. A¹ 208, fol. 32.

51. Girard d'Albissin, *Genèse de la frontière franco-belge*, pp. 114–18; Louvois to Charuel, 20 October 1667, A.G. A¹ 208, fol. 100.

intendants should have the lieutenant generals sign a large number of blank passports in advance and turn them over to the intendants to prevent their being farmed for the generals' benefit.[52]

Besides levying taxes, requisitioning forage, confiscating goods, and selling passports, the state decided to plunder the occupied area for seasoned wood suitable for artillery use during the next spring campaign. Early in October, shortly after the crown appointed the intendants, Louvois instructed them to search the forests in their departments. In a letter to Robert dated 8 October 1667, the secretary went into greater detail: the wood might be cut in forests belonging to subjects of the Spanish king and confiscated by right of war, or it might be obtained at low cost from the abbeys and other ecclesiastical lands in the area. Louvois particularly wanted Robert to search for large quantities of elm and oak to make gun carriages and thick planks for platforms.[53]

Louvois sent similar letters to other intendants. He asked Terruel to search in the wooded area of Mormal for timber, and suggested to Charuel that he investigate the forest of Nieppe. When Charuel complained that there was no properly seasoned wood in his department, Louvois suggested that if the wood were cut in October and allowed to dry until February, this would suit his purpose. The secretary also dispatched orders to intendant Talon to stockpile thick planks in the magazine at Audenarde so they could later be used for gun batteries. Charuel had to refit the artillery at Lille, where the war department wanted new mountings for twenty mortars capable of firing shot ten to twelve inches in diameter; "one must make them of such good wood and carefully bound with iron that they can resist the recoil of the mortar when it throws the bomb," Louvois wrote, "and not be like those this year that broke with each shot fired." Similarly, the army needed two hundred carts of various types and fifty covered wagons, all with iron-rimmed wheels. Intendant Terruel worked to supply sixty gun carriages for twenty-four-pounders, ten-pounders, and culverins, as well as twenty carts to carry the cannoneers. Louvois instructed him to try to obtain dry wood of good quality at whatever price he judged appropriate.[54]

52. Louvois to Charuel, 7 November 1667, A.G. A[1] 208, fol. 143.

53. Louvois to Robert, "intendant du costé de la mer," 8 October 1667, A.G. A[1] 208, fol. 64. Cf. Terruel's instructions, A.G. A[1] 208, fol. 32, which read in part: "Comme pendant l'hyver prochain Sa ma[te] pretend faire travailler aux esquipages d'artillerie de campagne, et que pour les faire meilleurs et a meilleur marché, il faut avoir des bois secqs, il s'informer bien soigneusement des lieux de son departement qui sont les plus propres et y faire travailler le bois des affuts."

54. Louvois to Terruel, 8 October 1667, A.G. A[1] 208, fol. 65; to Charuel, 8 and 14 October 1667, A.G. A[1] 208, fols. 62, 81; to Talon, 26 November 1667, A.G.

CARE OF THE TROOPS

Although none of the intendants of conquered areas bore the title of army intendant during the winter of 1667–68, as Louvois's trusted agents they could not help concerning themselves with military affairs. Their commissions justified their imposition of money and forage on the grounds that it was necessary for the subsistence of the large number of troops that would winter in the occupied territory. Yet there is little evidence that they played more than a minimal role in the supply of the army, which they left to the munitioneer and his assistants.[55]

Sometimes, however, Louvois asked an intendant to intervene. In October 1667, for example, Lieutenant General Lorges reconnoitered the area between the Sambre and Meuse rivers, and Louvois told him to obtain flour to make into bread from Charleroi and to transport it by boat, or to obtain it from the villages in the neutral bishopric of Liège. If there was a munitioneer's assistant accompanying the army, Louvois wrote, all Lorges had to do was give him the necessary orders. In case this agent was lacking, Louvois had instructed intendant Camus des Touches to meet with Lorges to see that the army was regularly supplied. In separate instructions to Camus des Touches, the secretary of war ordered him to use money collected from the contributions to handle this expense, promising the intendant that the munitioneer would reimburse him.[56]

This action took place in the last days of the autumn campaign, shortly before the army settled into winter quarters, and Camus des Touches's task remained exceptional. During the rest of the winter, while the army stayed in garrison, the correspondence of the intendants indicated little concern with feeding the army. Yet it seems that the intendant retained some jurisdiction over the distribution of forage, for an instance arose in February 1668 when the secretary of war had to write to Charuel about his reduction of forage for horses in winter quarters. Louvois explained that the cavalry had complained about the intendant's conduct. No doubt, the secretary assured him, there was some misunderstanding about the royal orders. Charuel had permitted only one ration of forage for each cavalier. This, in itself, was not unreasonable, but a cavalryman often had other "nags" (*bidets*) that he

A[1] 208, fol. 217; "Estat des munitions de guerre que le Roy a commandées pour les deux armées de Flandres," 6 December 1667, A.G. A[1] 208, fol. 284; Louvois to Terruel, 16 December 1667, A.G. A[1] 208, fol. 289.

55. See, for example, the commission of Carlier, 16 October 1667, A.G. A[1] 208, fol. 97.

56. Louvois to M. de Lorges, 6 October 1667, A.G. A[1] 208, fol. 46, and to Camus des Touches, same date, A.G. A[1] 208, fol. 49.

expected to feed with government assistance. The new policy was that the crown would furnish part of the additional forage, although the cavalryman with only one horse would continue to receive no more than one ration.[57]

The intendants of occupied territories also continued to exercise the army intendant's normal task of scrutinizing army finances. The army received income from two primary sources: contributions levied upon the conquered zone and sums sent separately by the government. In October, Louvois announced that he was dispatching three hundred thousand *écus* to Charuel for the subsistence of the soldiers lodged in Flanders; he added that he hoped to send an equal sum within a few weeks. That same month, the war secretary forwarded a bill of exchange to Robert for the payment of the troops. Sometimes the war department provided money for other military purposes, such as in December 1667 when Louvois transmitted thirty thousand *livres* to Charuel for the construction of boats for pontoon bridges.[58]

Naturally the intendant had to maintain accounts, for both the receipts and expenses of the occupation, and Louvois did not hesitate to demand either. In late September 1667, Louvois requested a statement of expenses for the subsistence of the troops in Charuel's department so the war secretary could make an accounting to the king. When an intendant was lax in submitting his figures for auditing, Louvois firmly reminded him of his duty. Thus he wrote to Robert in late November 1667: "I believe I told you that at the end of each month you must send me the statements of receipts and expenses, both for the extraordinary funds as well as those from the contributions." [59]

Often when there was a need for money and the funds of the war treasury were temporarily exhausted, the intendant had to dip into funds from the contributions—though only with the government's permission. Sometimes this amounted to little more than juggling funds or a sleight of hand. This accounts for an incident that took place in early October 1667: Charuel wrote to Louvois that he had given Monsieur Faille, director of the army hospital, authority to deduct fifteen thousand *livres* from the contributions of Courtrai for support of the hospital, as requested by intendant Colbert. When Louvois sent his promised thirty thousand *livres*, the intendant informed the secretary, he planned on using fifteen thousand of it to repay the receiver of

57. Louvois to Charuel, 2 February 1668, A.G. A¹ 222, fol. 198.
58. Louvois to Charuel, 8 October 1667, A.G. A¹ 208, fol. 62; to Robert, 22 October 1667, A.G. A¹ 208, fol. 112; to Charuel, 16 December 1667, A.G. A¹ 208, fol. 283.
59. Louvois to Charuel, 23 September 1667, A.G. A¹ 208, fol. 17, and to Robert, 27 November 1667, A.G. A¹ 208, fol. 220*bis*.

contributions at Courtrai. To give the transaction legality, Charuel arranged for Monsieur Faille to give the receiver of Courtrai a receipt that he, in turn, would present to the war treasurer for reimbursement. In another example, Louvois asked Terruel to pay the soldiers from the funds of the contributions or by his own credit until money arrived from the government. The point is clear: the state did not hesitate to use its intendants in financial manipulation if this served its purpose.[60]

The same intendants had the responsibility of investigating any fiscal irregularity in the conquered zone. French occupation had barely begun in 1667 when Louvois wrote to Charuel that Deslandes, governor at Ath, had complained that there was no money for work on the fortifications of the town. This surprised the secretary, but he ordered Charuel to deliver additional funds. Meanwhile, the intendant was to "watch these works and the manner in which the expenses are made, so that the king's money is not spent without necessity and that no abuse is committed there." Similarly, Louvois wrote to Robert in mid-November commenting that he had received his statement on the expenditure of the four thousand *livres* annually raised in the castellany of Bourbourg for the maintenance of the barracks at Gravelines. "I have always been convinced," Louvois stated, "that there was some tomfoolery [*grimelinerie*] there, and I am made even more sure by the reading of your accounts. Please investigate and tell me what it is. Make clear to those who receive this sum that it must not be dispensed without your orders." [61]

Since intendants investigated the behavior of others, it was important that their own conduct be irreproachable. Only rarely did a case occur when the war secretary suspected that he had misplaced his trust. In January 1668, Louvois had doubts about Talon, intendant at Audenarde. Not only did it appear that the intendant's affairs were in confusion, but Talon had not kept the secretary informed of activity

60. Charuel to Louvois, 8 October 1667, A.G. A¹ 209, fol. 266, and to Terruel, 29 October 1667, A.G. A¹ 208, fol. 126. It should be noted that the government was devious in other matters. For example, Louvois wrote to Charuel, 10 November 1667, A.G. A¹ 208, fol. 144, that he and the military commander, Monsieur d'Humières, should tolerate the soldiers raiding the forests adjacent to the places in which they were garrisoned for fuel to heat themselves. They should make it appear to the inhabitants, however, that no consent had been given and, indeed, that the government was doing all in its power to prevent the illegal robbing of the wooded areas.

61. Louvois to Charuel, 8 October 1667, A.G. A¹ 208, fol. 62, and to Robert, 16 November 1667, A.G. A¹ 208, fol. 176. In a third incident, Louvois wrote to Robert, 31 October 1667, A.G. A¹ 208, fol. 134, asking him to examine the conduct of an official at Bergues named Dazin whom Louvois suspected of peculation, for despite Dazin's great expense he had not made the castellany secure.

in his department, particularly in regard to monetary matters. Louvois asked Charuel what he knew of Talon's conduct, and ordered him to question secretly the receiver of contributions and any others who might know about the intendant's behavior. Furthermore, the war minister did not believe that the *commissaire des guerres* who served under Talon was very careful (*bien seur*) with money, although he had a long record of military service. Louvois inquired if Charuel knew anyone in his own department who could be trusted to conduct an investigation without arousing suspicion. All other letters concerning this inquiry are missing from the war archives, but it can be concluded that Charuel silenced his master's suspicions, for Talon retained his office of intendant. The incident underscores the secretary's reliance upon the intendant as his most trusted agent and at the same time the importance of his probity.[62]

HOSPITALS

Another immediate concern of the intendants of contributions remained the care of sick soldiers. These patients included not only the casualties from the summer campaign, but also those who fell ill while stationed in the garrisons. When a number of soldiers suffered from sickness in November 1667, Charuel explained the cause of their malady: the poor condition of the barracks, the lack of homes in which to quarter the soldiers, and the haste in obtaining lodging, particularly at Armentières. But Charuel saw signs of hope, for there were not as many maladies at Lille as at Armentières, as a result of the excellent care by the nuns at the nearby hospital of Seclin. In return for this care, Charuel had exempted the convent from the French impositions, and he remarked that it deserved the protection of the king.[63]

Not all the sickness that occurred during the winter can be explained by the poor housing. Contagious disease always seemed to follow armies, and that in Flanders was no exception. An epidemic variously called "the pest," "the contagion," or "the malady" quickly spread

62. Louvois to Charuel, 6 January 1668, A.G. A¹ 222, fol. 16. In spite of Louvois's comments regarding the *commissaire des guerres*, his suspicions centered around Talon because, as the secretary remarked in this letter, "le peu desclaircissem^{ts} quil ma donné jusques a present sur le destail des affaires de son departement me faisant soubçonner au dernier point de sa netteté a l'esgard de l'argent." I carefully searched the archives but could not find any other letters referring to this incident.

63. Charuel to Louvois, 22 November 1667, A.G. A¹ 210, fol. 19. The religious hospital at Seclin was a famous charitable institution. Founded in the thirteenth century by Marguerite of Flanders, it continued to grow in size during the fifteenth and seventeenth centuries.

through the occupied towns. On 17 and 18 November 1667, Charuel reported, twenty-five persons had died in the city of Lille, another thirty were dead in the hospital, while twenty-nine others had fallen ill. On 19 November, he continued, the contagion killed another sixteen persons; 20 November saw twelve more dead, and by the twenty-first the figure had increased by twenty-four fatalities. The disease spread to Courtrai. At Audenarde, Talon recorded the number of sick at five hundred and remarked: "we must prevent it [the disease] by any means possible from reaching the troops, for it attacked the Spanish at Cambrai and all perished."

Rumor reported that the plague had spread to Ath, while there were forty infected houses in the Spanish city of Ghent. The plague ran its course in most of the French territory by January, when the freezing weather put an end to the disease, but Camus des Touches was swamped at Charleroi in midwinter trying to care for the vast number of maladies. He had established a hospital in the town, but as he lacked medications, he wrote to his brother Camus du Clos to purchase some three hundred *livres* of supplies at Charleville and to obtain a friar from the order of La Charité to take charge of the hospital.[64]

The efforts of the intendants were limited in the face of this scourge. Gaboury recognized that overcrowded conditions facilitated the spread of the disease among the troops at Saint-Venant, and he conferred with the governor about the possibility of building huts on the town ramparts to house the soldiers until the plague subsided. Despite his efforts, the pest continued its rampage. Given the primitive medical facilities of the time, the intendants and other civilian authorities tried to obtain relief by such crude sanitation efforts as burning the bedding of those who had died and stationing soldiers at the city gates to prevent the entrance of inhabitants from infected areas. But in reality the intendants could do little.[65]

FORTIFICATIONS

As soon as the French armies occupied the Flemish frontier, the intendants began examining the captured strongholds to see what work

64. Charuel to Louvois, 19, 22, and 25 November 1667, A.G. A¹ 210, fols. 7, 19, 35; Talon to Louvois, 23 and 25 November 1667, A.G. A¹ 210, fols. 22, 23; Camus des Touches to Louvois, 2 January 1668, A.G. A¹ 224, no. 5.

65. Gaboury to Louvois, 10 and 14 March 1668, A.G. A¹ 225, nos. 60, 88; Louvois to Gaboury, 26 January 1668, A.G. A¹ 222, fol. 139. It is difficult to define the contagious disease mentioned in the above two paragraphs. Gaboury called it "la peste," Louvois spoke of the soldiers "morte d'une maladie que l'on soubçonne de peste," while Charuel spoke of people "mortes de contagion."

needed to be done to ready the place for the next campaign. The crown decided to dismantle some fortifications so they would not become a dangerous threat if they were reoccupied by the enemy, and it hoped to shore up other fortresses so they would resist future attacks by the Spanish. In mid-October 1667, Louvois announced to Charuel that the king had approved Vauban's plans for the fortress of Courtrai. His Majesty desired that the intendant pay for the necessary expenses involved in putting the plans into effect. The project called for the removal of a large quantity of earth, the placement of palisades, and the planting of sod. Louvois suggested that the best way for Charuel to obtain labor for this scheme would be to demand workers from the surrounding area by the month, that is, calling for six to eight men for labor service from the larger villages and at least three or four from the smaller. This, the secretary thought, would be the cheapest way, and he admonished the intendant "to manage the money of the king, so that it does not cost His Majesty much and does not exhaust the countryside." [66]

Although Charuel was not the only intendant to work on fortifications—for example, Camus des Touches worked at Charleroi and Terruel at Quesnoy—the bulk of the labor fell upon his shoulders. This was probably because Louvois trusted him and also because many of the fortresses in occupied Flanders were in his intendancy. Charuel worked at other fortifications besides Courtrai, namely Lille—the intendant's place of residence—the fort at Douai, the citadel of Tournai, and the palisading at Audenarde, Ath, and Armentières. Although he shared responsibility with the military governor (Humières), the receiver of contributions (Pertuis), and the engineer (Vauban), Charuel supervised many details of the work, such as procuring a contractor for skilled labor, searching for suitable timber for the palisades, and reporting to Louvois.[67]

66. Louvois to Charuel, 14 October 1667, A.G. A¹ 208, fol. 78.
67. Louvois to Charuel, 13 and 20 November 1667, A.G. A¹ 208, fols. 160, 200. See also Louvois to Camus des Touches, 26 December 1667, A.G. A¹ 208, fol. 337, and to Terruel, 27 December 1667, A.G. A¹ 208, fol. 343. One cannot over-emphasize Louvois's confidence in Charuel. For example, the secretary trusted him with classified information. He revealed to the intendant (in the letter of 13 November) that the king had decided to raze the fortress of Armentières the next spring, but that the secret was for Charuel's ears alone: "mais que cela soit pour vous seul, et que l'on n'en entendre jamais parler." Curiously, Louvois also cautioned the intendant to watch Vauban, in a letter dated 14 October 1667, A.G. A¹ 208, fol. 78: "le Sʳ Vauban est asseurement capable de bien servir, mais il n'est pas inutile de l'exiter a bien faire vous luy tesmoignerez qu'il doibt mettre en pratique son industrie pour faire faire [*sic*] les ouvrages a bon marché et tres promptement afin que lon puisse faire voir au Roy que les mauvais offices qu'on luy a rendus sur cela sont mal fondées." Perhaps this refers to the following inci-

RELATIONS WITH ARMY OFFICERS

Regular army intendants, accompanying the troops in the field, closely cooperated with the commanding generals. Similarly, the intendants of the conquered zone maintained contact with the military commanders in the frontier region. Often the letters of Louvois contained the same information for the intendants and the officers. In separate correspondence to Carlier and Créquy, dated 20 November 1667, Louvois informed both men that the king had decided to maintain Créquy's two aides-de-camp during winter quarters, and he ordered Carlier to furnish their forage on the same basis as a captain of light cavalry. In the same letter the secretary informed the intendant and the general that the infantry in the future would receive five *sols* per day because the army preferred its entire pay, rather than four *sols* and a ration of bread. The secretary also reported that the king had approved the purchase of gun powder from the family of the late Marshal Fabert at the rate of seven *sols* a pound and no more. Louvois told Créquy that Carlier would conclude the contract "under your orders," and he wrote the intendant that he should buy the powder in union with Créquy.[68]

Although the intendant and the military governor were the two key officials in the occupied zone, it is apparent to anyone who has examined Louvois's correspondence that he often placed his confidence in the intendant rather than the governor. A letter of Louvois to Charuel, dated 8 October 1667, transmitted a set of instructions with the injunction that he explain them to the general: "to avoid the effort of making a grand recitation to Monsieur the Marquis d'Humières, which would consume much of my time and which he, no doubt, would understand much less than a conversation with you, I have notified him that you will communicate everything that concerns the king's service at Lille and in your employ. And I must tell you that it is apropos for you to explain it to him carefully [*fort exactement*]." In a letter in November, Louvois asked Charuel to put a receiver of contributions in charge of the construction workers at Courtrai. Otherwise, the secretary warned,

dent: Vauban had been unjustly implicated in a swindle by Colbert de Saint-Marc, a cousin of Colbert and successor of Colbert de Croissy as intendant of Alsace, who had falsified the accounts on subcontracting work done on the fortifications. The case went to court, and Vauban's reputation was not finally cleared until 1671. See Michael Parent and Jacques Verroust's brief account, *Vauban* (Paris, 1971), p. 64.

68. Louvois to Créquy and to Carlier, both 20 November 1667, A.G. A¹ 208, fols. 197, 198. Cf. also Louvois to Humières and to Charuel, also dated 20 November 1667, A.G. A¹ 208, fols. 199, 200, instructing them to act "en concert."

a bourgeois of the town would have the responsibility and "would depend a little too much on the governor." [69]

Louvois, of course, always wished that an intendant would respect a military commandant's rank. Thus before the marquis de Bellefonds arrived to take command of the troops wintering between the Sambre and Meuse rivers, the secretary asked Camus des Touches to provide an escort of cavalry and to visit the general to inform him of the details of the contributions and the lodging of troops. Be on good terms with the marquis, Louvois admonished, and give him all the respect possible. "If it happens that in working with him on the king's affairs, he does not have the same sentiment as you, and even if he desires some change in your actions, you must softly recite your reasons, but, nevertheless, conform to his sentiments. Never oppose him directly, but because you must make a report here of all that occurs in your employ, you will be careful to let me know about everything that passes between you and him, and particularly what is contrary or different from the orders and intentions of His Majesty." [70]

Sometimes it is very difficult to tell when this careful surveillance crossed the nebulous borderline and became spying. Louvois wrote to Terruel to "inform me, even to the least detail, of all that occurs in your department." The secretary expressed the same idea in a letter to Carlier: "You will carefully observe the conduct of those who have authority over the troops, [those] who command in the places of your department, and those who serve under them. In case some of them contravene His Majesty's intentions, you will give immediate notice here." In such violations, the intendant need not observe distinction with respect to rank. "Please remember and put it into your head once and for all," Louvois scolded Camus des Touches when the latter failed to keep him informed of the instructions given by the count de Broglio to palisade the town of Avesnes, "that you must make no distinction of persons in the execution of the affairs regarding your employ." [71]

It would be too easy to say that Louvois placed his agents to spy upon the military. It is wiser to regard the intendants as troubleshooters placed in strategic areas. Sometimes the intendants occupied themselves with the delicate task of reconciling differences between various com-

---

69. Louvois to Charuel, 8 October and 18 November 1667, A.G. A^1 208, fols. 62, 190.

70. Louvois to Camus des Touches, 9 October 1667, A.G. A^1 208, fol. 66.

71. Louvois wrote to Terruel, 29 October 1667, A.G. A^1 208, fol. 126, mentioning that he should continue "m'informer jusques aux moidres choses de tout ce qui se passera dans vostre departement." Louvois to Carlier, 18 October 1667, A.G. A^1 208, fol. 95, and to Camus des Touches, 27 September 1667, A.G. A^1 208, fol. 28.

manders, so that the government's plans did not flounder on the rocks of personal antagonism. Early in January 1668, Charuel detected a sensitive spot in the count de Duras's behavior; the intendant had witnessed the count's displeasure when royal letters were sent without his knowledge to various military governors under his jurisdiction. Charuel wrote to Louvois that this slight tormented Duras because he was very jealous of his authority; moreover, the count already had differences with certain governors because he tried to intervene in the affairs of their areas.

Due to the intendant's perceptiveness, Louvois was able to soothe the count's ruffled feelings. If, Louvois informed Duras, he had unintentionally omitted anything from his conduct, he would correct it immediately. He explained that when he had written to the local governors, it was at the king's desire. Even then, he had tried to make Duras aware of the particulars. He concluded the letter by asking what he could do to please the count. Duras, undoubtedly pleased by the suppliant tone of the minister, muttered that he wanted to let the matter drop, as he was certain that the governors would be more careful of his position in the future. It is obvious that unless an intendant was on hand to warn the secretary about sensitive issues, sharp disagreement could arise between army and administration. In this instance, an intendant mediated between the war department and its army officers.[72]

JUSTICE

In matters of justice, the intendants of contributions had no legal authority whatsoever—their commissions made no mention of any judicial function. In occupied areas the government normally tried to leave law and order in the hands of the legally constituted native officials. Only when these men did not or could not fulfill their responsibility did the intendant step in. Thus Louvois asked Charuel to warn the magistrates at Lille that they must prevent the inhabitants from insulting the French soldiers stationed there. If the local officials did not take appropriate measures to end this abuse, General Humières would

72. Charuel to Louvois, 3 January 1668, A.G. A¹ 224, no. 9. Louvois in his letter to Duras, 6 January 1668, A.G. A¹ 222, fol. 17, explained: "M. Charuel m'avait desja mandé par advance qu'il croyoit que vous vous plainderiez de ce que vous n'aviez pas de part dans tout ce qui se fait dans vostre département, comme je crois que vous ne pouvez pas douter de ce qui me regarde, je luy ay mandé de vous faire explicquer la dessus afin que si sans le sçavoir et sans en avoir intention, j'avois obmis de faire quelque chose de ce que vous desirez de mon service je me corigeasse a l'avenir." Duras's reply to Louvois is dated 13 January 1668, A.G. A¹ 224, no. 91.

take matters into his own hands and immediately hang the first man caught in the act. Meanwhile, during Camus des Touches's intendancy, Louvois simply told the intendant to order the provosts in the infantry regiments to condemn (*jugez*) the thieving peasantry of the area.[73]

In December 1667, the government decided to regulate justice in a more permanent manner in the Flemish territories. The crown planned on setting up a court of appellate jurisdiction to handle appeals from local courts. Since the king needed to appoint two men of integrity from each of the towns of Lille, Tournai, Douai, and Courtrai, plus one man each from Audenarde and Armentières, Louvois asked Charuel to nominate appropriate local officials for the bench. The secretary admonished Charuel not to tell anyone of his choice, for the king might decide to allow the cities to choose their own candidates, with the exception of a crown-appointed president. The conclusion is apparent: the intendants of contributions in the occupied areas were administrators who left justice to the native courts. When they intervened, it was not as judge but as bureaucrat.[74]

Even in the army camps, the correspondence confirms the fact that the intendants of contributions played a minimal role in the administration of military justice. January 1668 seemed to be a month of particular dissatisfaction and lawlessness among the soldiers stationed in Flanders. On 8 January, Talon reported from Courtrai that a sutler, one who sold provisions to soldiers, had been condemned to be flogged and branded with the fleur-de-lis. The three cavaliers who assisted him were to be present at the execution of the sentence (*d'y assister en entrer après cela dans le régiment*). Talon had already sent for an executioner to come from Tournai to administer the punishment when the count de Duras suspended the sentence. Talon complained that the men had pillaged a wagon and that rigorous administration of justice was needed to maintain discipline, so he worried about the effect of the suspension.

Duras justified his action to Louvois by claiming that the soldiers had not been tried according to the proper forms and that they had paid compensation to the owner. Moreover, they had remained in prison for over a month. He felt this was punishment enough, and upon the recommendation of his officers, he had returned the culprits to their companies. Louvois, a wise administrator, knew that it was judicious to support the authority of a general, and he announced that the king had approved the lifting of the sentence "in consideration of M. de Duras," but that henceforth an example was to be made of all crimes

73. Louvois to Charuel, 10 November 1667, A.G. A[1] 208, fol. 144; to Camus des Touches, 14 November 1667, A.G. A[1] 208, fol. 164.
74. Louvois to Charuel, 7 December 1667, A.G. A[1] 208, fol. 239.

in order to maintain discipline among the troops. Thus both the intendant and the general managed to save face.[75]

During that same January, a minor sedition broke out among the troops stationed at Bergues, a town in Robert's intendancy. Robert was in residence at Dunkirk, so he played no role in the arrest and trial of the mutineers except to report the incident to his master. Some twenty-five or thirty cavaliers of the company of Beaupré, discontent because they had not received their pay, stormed off to see their commanding officer, Du Passage, who was visiting at the home of the burgomaster of Bergues. The troops assembled in front of the house and demanded that the commander make an appearance. Du Passage acted decisively, seizing the spokesman and demanding that the officers of the company arrest the most unruly of the soldiers. Cazeux, the military governor of the town, presided over the council of war acting as a court, which found extenuating circumstances due to the fact that the captain had withheld money from the soldiers' pay in order to buy clothing for the spring campaign.

Although Robert played no role in the trial, he tartly remarked: "Even if the officer had done a greater wrong than he had, the cavaliers were nonetheless guilty, for nothing in the world permits soldiers to go en masse to a general's residence to complain." The intendant had no doubt that Louvois would reverse the court's clemency; he was not disappointed, for the secretary sent a blistering letter to Du Passage in which he attributed the court's mercy to its inexperience and ordered the council of war to be recalled into session to reconsider its verdict. This time, Louvois maintained, the council should impose the death penalty upon the ringleader and lead him to the place of execution. But Louvois commanded it to be announced at the last moment that the king had pardoned him. Robert took no part in all this; he acted merely as an adviser who kept the king informed of events in his department.[76]

75. Talon to Louvois, 8 January 1668, A.G. A¹ 224, no. 58; Duras to Louvois, 4 January 1668, A.G. A¹ 224, no. 14; Louvois to Duras and to Talon, both 18 January 1668, A.G. A¹ 222, fols. 97, 102. Actually there was a slight discrepancy in the stories of Talon and Duras. Talon described the sutler as a *vivandier soldat*, but Duras said the man was "un soldat de qui la femme sert de vivandiere." It seems the wagon owner was trying to sell the soldiers wood for 3.5 *sols*, but they valued the wood at only 2 *sols*, and a brawl ensued.

76. This and the above paragraph are taken from the following correspondence: Cazeux to Louvois, 4 January 1668, A.G. A¹ 224, no. 20; Robert to Louvois, same date, A.G. A¹ 224, no. 21; Du Passage to Louvois, 5 January 1668, A.G. A¹ 224, no. 22; Louvois to Robert, 17 January 1668, A.G. A¹ 222, fol. 94; Louvois to Du Passage, 18 January 1668, A.G. A¹ 222, fol. 105.

Epilogue

By 1670 the dimensions of certain long-range trends in the office of army intendant can be gauged. Although the origin of these officials remains uncertain, they developed amid the army camps in the latter half of the sixteenth and early seventeenth centuries. The office evolved from special inspectors or commissioners with unlimited powers, comparable to other civilian officials on mission inside France such as treasurers, financial experts, masters of requests, members of the sovereign courts, and councilors of state.

The primary function of the first intendancies was to administer justice amid the excesses of the troops and to punish rebellion against royal authority, and there was often a curious mixture of civilian-military activity in the performance of duty. Consequently it is difficult to distinguish between the early provincial and army intendants. The provincial intendants made a successful transition from the army camps to the largest city of their department. There they stationed themselves as permanent officials, although they continued to be concerned with some military matters such as recruitment, lodging, and supply of soldiers stationed in their locality. Unlike the provincial intendancies, those in the army appeared to remain vestigial. Although the crown often employed the same personnel and renewed their commissions, the office remained a specific assignment for the duration of an army campaign. Almost continual warfare during the administrations of Richelieu and Mazarin provided the reason for its survival.

Since the office remained flexible, it adapted to changing conditions—internal rebellion, foreign expeditions, and administration of conquered territory—during the first three-quarters of the seventeenth century. A broad, heterogeneous personnel occupied the army intendancies. In the 1630s and 1640s provincial intendants often accepted additional commissions for the armies stationed in their departments, or bright young masters of requests on their way up the governmental ladder used an army intendancy as a stepping-stone to a more important position—Michel Le Tellier is the best example. For others such as Robert Arnauld d'Andilly, the commission was a capstone to a career.

In brief, a wide variety of officials occupied military intendancies according to the exigencies of the moment. During the time of the Fronde, the crown used army intendants as crypto–provincial intendants, often, but not always, the very same officials whose positions had been abolished. This period also witnessed an attempt by generals to take advantage of the weakness of the crown to influence the choice of personnel. But with the passage of time and the rise of Le Tellier, the army intendants developed into a permanent group of clients around the war minister, whom they served in periods of war and peace as army intendants or as intendants of occupied territory.

There remained a dichotomy in the personnel. Sons from traditional high-officeholding families acquired a mastership of requests and the necessary legal education for entrance into royal administration. There they served as occasional army and provincial intendants before ending up as diplomats or ministers of state. The variety of experience and service in different departments provided the expertise they would need in the future. At the same time, a lower rank of the officeholding class often gained a foothold in the royal administration due to their army intendancies. These were the clients who worked in the war department as *commissaires des guerres* or *commis*. They did not possess a mastership of requests but were schooled by experience. They underwent a period of testing and training to prove their administrative abilities, first as *commissaires*, then as intendants in minor campaigns. After proving themselves, they provided long years of service to Le Tellier and Louvois.

The career of Jacques Charuel is almost a classic example. After accompanying a naval expedition to North Africa in 1664 and serving as an intendant of conquered territory during the War of Devolution, he acted as intendant of Créquy's troops during the occupation of Lorraine in 1670. He served once again as army intendant during the first two years of the Dutch War until he clashed with Turenne. After this incident, Louvois shifted him to the intendancy of the frontier province of Lorraine, where he remained until his death in 1691. His career illustrates that a mastership of requests was not the only entrance to an important government position, although the fact that Lorraine was a frontier province under Louvois's jursidiction implies that regular provincial intendancies remained the prerogative of the masters of requests.

The apprenticeship system still operated in 1670, the final year of this study. Although there were no major conflicts during the years immediately preceding the Dutch War, minor military action did occur. Such action gave two raw, inexperienced men a test under battle conditions, allowing them to prove their merit while facing the pressure of combat. Both were relatives of Louvois: the sieur de la Croix and Gilbert Colbert, sieur de Saint Pouenges. Both passed the test and seemed to have a brilliant future before them. Death cut short one man's career, but the other survived to become one of Louvois's most trusted advisers in the war department.[1]

Croix accompanied an expedition to Crete in 1669 as army intendant. As earlier during Robert's first intendancy, the war department did not overburden the young man with the responsibility for the campaign. It sent François Jacquier, an experienced munitioneer, as superintendent and commissary general of supplies to assist Croix. Unfortunately, Croix died of illness on the voyage home to France. Louvois acknowledged the news of the death of his protégé by lamenting to the duke of Navailles, the commander of the expedition: "I have taken the death of M. de la Croix with acute grief [*un sensible déplaisir*], for by your account of the manner in which he conducted himself since he has been with you, he had given me reason to hope that he would have served the king usefully in the employment that would have been conferred upon him in the future." [2]

If Croix's training came to a tragic end, Louvois found success with

1. I have not been able to discover the exact relationship of Croix to the Le Tellier family, or even his proper name. André, *Michel Le Tellier et l'armée*, p. 636, describes him as "un cousin." Before his first intendancy, Croix worked under Charuel's authority at Tournai, where he supervised the fortifications of that town; see his letter to Louvois, 1 December 1667, A.G. A^1 210, fol. 66. See also his letter of 29 June 1668 to Louvois, A.G. A^1 227, fol. 188, in which he complains of Le Peletier de Souzy's arrogance upon assuming the intendancy in Flanders: "tant que nous fusmes seules la conversaᵒⁿ ne fut que de choses indifferentes, et du moment qu'il vit sa chambre pleine de monde, il prit un tres grand soing de me faire connoistre a tous propos que j'estois de sa dependance." Concerning Saint Pouenges, see B.N. *dossiers bleus* 203, family "Colbert," fols. 1–27. He too had served as a *commis* in the war department before his intendancy. See Louvois to M. de Montal, 21 November 1667, A.G. A^1 208, fol. 203, regarding one of Saint Pouenges's missions.

2. Jacquier's commission, 15 March 1669, A.G. A^1 238, pp. 6r–8r; Croix's commission, 6 April 1669, A.G. A^1 238, pp. 33v–36v. For brief biographical and bibliographical information on Jacquier, see Julian Dent, "An Aspect of the Crisis of the Seventeenth Century: The Collapse of the Financial Administration of the French Monarchy (1653–61)," *Economic History Review*, 2d ser., 20 (1967): 248n5.

Louvois to Navailles, 11 October 1669, A.G. A^1 238, pp. 96v–98v. Navailles had reported Croix's death in an earlier letter to Louvois, 5 October 1669, A.G. A^1 238, no. 122: "J'ay un extreme desplaisir d'estre obligé de vous faire sçavoir la mort de Mr de la Croix nostre intendant, je suis asseuré que vous en serez sensiblement touché; mais je vous asseure que vous ne le sçauriez estre plus que je le suis, il est mort dans nostre passage proche le golfe de Palme d'un transport au cerveau qui la emporté en deux fois 24. heures,"

Gilbert de Saint Pouenges. His apprenticeship began in 1670, when the king named him as army intendant for the expedition to occupy Lorraine. There he served from August until the beginning of winter quarters in October, when Louvois replaced him with the older professional Jacques Charuel. Nevertheless, the secretary was satisfied with Saint Pouenges's behavior, for he quickly rose in the war department. Rather than service in the field as army intendant, his talent lay in desk work. By 1673 Louvois had given him permission to open letters during the secretary's absence and report the information to the king. He became one of the secretary's closest advisers and ended his career with the honorary position of grand treasurer of the king's orders.[3]

The war department's training program continued after 1670. The career of Germain-Michel Camus de Beaulieu illustrates this process. He and his brothers had accompanied Robert as *commissaires des guerres* during the expedition to Hungary in 1664. He continued to serve as military commissioner during the War of Devolution and assisted in preparing for the Dutch War. In 1673 he replaced Charuel as intendant of Turenne's army after Charuel's difference with the general. He later worked as intendant of the army of Roussillon (1675) before becoming provincial intendant of that region the following year. He remained in Roussillon until 1681, when he obtained the position of controller general of the artillery.

A fundamental change in an army intendant's function assisted former *commissaires* to ascend to that rank. No longer were army intendants judicial officials as they had been in the days of Laffemas and Machaut during the difficulties of the 1630s. As the years passed, the function of justice atrophied, while the essentially administrative functions of police and finance increased. The army usually left justice to the provosts and *conseils de guerre* sitting as military courts; the army intendants usually did little more than report the criminal situation to the government. Therefore, when cases occurred in the 1670s, an army intendant might not know how to act.

The best illustration of this is an incident during the Dutch War. Du Pas, an army officer, had made a feeble defense of the town of Naerden, surrendering it to the enemy after only four days' resistance. Louis XIV, infuriated, ordered the commanding general, the duke of Luxembourg, the intendant, Robert, and a number of high-ranking army officers to conduct a trial. Robert, uncertain over his role, anx-

3. In 1673, during Louvois's absence in Alsace, Saint Pouenges acted as his go-between with the king and "rendoit compte au Roi de ce qui concernait le ministre." See Saint Pouenges's letters during August, A.G. A^1 316, especially pp. 122r–28r.

iously wrote Louvois for further instructions. He planned on using a *commissaire* named Sauvé, who understood legal procedure, as prosecuting attorney (*procureur du roy et lieutenant de la connestablie*). "For myself," Robert remarked, "I do not understand these sorts of things," and he asked the secretary to instruct him on his conduct. Thus by this time intendants had become administrators, powerful bureaucrats with a flock of underlings—men who owed their position to their executive talents and their devotion to the war secretary. An intendant's experience in handling financial matters, maintaining supplies, and keeping the secretary constantly informed mattered more than legal expertise.[4]

The army intendant remained an important agent of royal power in the army, but he was in a supervisory capacity as chief of the civilian-military hierarchy. In practice, he delegated much of the minutiae of administration to his ever-present subordinates, the *commissaires des guerres*. They were the officers who dealt with local officials, purchased needed supplies, supervised the hospitals, and attended to a hundred other details, but the intendant remained their superior.

Although the *commissaires* accepted this superiority, the army intendants continued to have difficulty in asserting their influence against the officer class, composed of aristocrats who hoped to return to an earlier period when they had made the army their private preserve. As this study has illustrated, the war department tried to exercise civilian control over these officers, and frequent clashes broke out between the army commanders and the intendants, such as between Robert and Coligny. Powerful generals like Turenne might force the removal of intendants like Charuel, but Charuel was not disgraced; he was promoted to provincial intendant of Lorraine.

An incident in 1670 illustrates this theme: the army intendant's real role and function was to assert the crown's authority. During the occupation of Lorraine, Louvois replaced the novice Saint Pouenges with the experienced Charuel, who had instructions to order everything necessary for winter quarters in his own name rather than the general's. At first Marshal Créquy contented himself with remarking: "In the employs with which His Majesty has honored me until the present, he

4. The king to Robert, 21 September 1673, A.G. A¹ 339, fol. 244; Robert's reply to Louvois, 2 October 1673, A.G. A¹ 339, fol. 270, in which Robert commented: "Car pour moy je n'entends rien du tout a ces sortes d'affaires, et mesme je ne sçay pas lors qu'on jugera l'affaire, si ce debura estre M Sauvé comme procureur du roi qui la raporte dans le conseil de guerre, ou si ce debura estre moy qui la raporte, et dire en mesme temps mon advis ou si je ne deburay point la raporter, mais seulement dire mon advis comme les autres juges, et en ce cas la, en que rang je deburay dire mon advis, je vous supplie tres humblement de me faire instruire de cela."

has done me the favor of giving me the direction of these things, and it will be in his best interest if the intendants have the same conduct as in the past. . . . Otherwise, Sire, one would scarcely place any credence in a general."

When this protest had little effect, Créquy angrily denounced the intendant: "Until Charuel's arrival, I had believed that the king's intention was that he [Charuel] would plan the impositions for the countryside, that he would make the levies of the moneys ordered by His Majesty, and that he would enter into the details of the distribution that must be made to the troops—all this by orders [*les envoyes*] of His Majesty, signed by me and countersigned by him. However, I quickly learned that he [Charuel] wanted to order [*ordonner*] everything in his own name and that he claimed that everything to be done in this country should be decided, regulated, and done under his pleasure." This annoyed Créquy, and he remarked that when it had pleased the king to employ him in Luxembourg and elsewhere, His Majesty "always left me the authority that one should have over an intendant." If the king still wanted things done in Charuel's name, Créquy assured the monarch, the general would conform to his wishes, but the king would be better served by supporting the general's authority than by leaving the final decision to Charuel.[5]

The government did not grant Créquy's request. It had decided that it was the general's prerogative to have the final authority during the campaign. But when the army had concluded the active phase of fighting and entered into winter quarters or the occupation of territory, it was the intendant who was to have the final disposition of things. He was to consult the general and listen to his advice, but he did not need to have the general's formal consent. The intendant had risen to partial equality with the general.

Civilians gained this power partly because they were more efficient administrators, but also because the government did not trust its military officers to levy impositions, fix rations, distribute pay, or arrange lodgings. There were too many opportunities for pecuniary gain and financial fraud. Moreover, the crown did not want the generals to use occupied territory as a power base where they could control patronage and make themselves independent. Charuel's instructions were one more attempt in a series to trim the claws of the powerful military. Of course the generals violently objected to their loss of power, and Créquy blustered: "To tell the truth, I do not believe that with an

5. Créquy to Louvois, 12 and 25 October, A.G. A¹ 250, pp. 120v–24v, 155r–57r. Relevant portions of these letters are in Camille Rousset, *Histoire de Louvois et de son administration politique et militaire* (Paris, 1862–63), 1:304, 306–7.

army of twenty thousand men I should be treated on the same basis as a governor of a province." But this was exactly what the crown desired, and it used the army intendants as the instruments of its will.[6]

An intendant's power rested on two foundations: his administrative expertise, and the support he received from the king and the secretary of war. The first is obvious; the second should not be underestimated. The war department supported Charuel at every moment in the struggle with Créquy, and with the government's authority behind him it would have taken an inept intendant, something that Charuel certainly was not, to throw away this advantage. As long as the intendants remained the tool of royal authority, their power could only increase.

It would be simplistic to claim that there was a vicious power struggle between the commanders and the army intendants. It was more like a slow, gradual process by which the intendants with the secretary's approval, if not prodding, gained civilian control. The process had not ended by 1670, yet the trend was apparent. But the balance between civilian and military always remained a delicate one: there had to be harmony and cooperation between the two spheres or the monarchy's military designs suffered. Consequently Louvois always urged the intendants to treat the generals with the respect that their position merited, to consult with them, and to listen to their advice. The war department desired that the generals do the same. Fortunately for the royal service, most generals, beginning with Brézé in the 1630s, reached a modus vivendi with their intendants and worked closely with them. Nevertheless, the situation remained sensitive, and the government feared it would upset the status quo if it acted too abruptly.

Part of the explanation for this sensitive relationship remained the dual nature of an intendant's function. In part, he was an able administrator who handled matters of finance and supply, and as such his position was indispensable—even generals like Créquy or Coligny recognized this. Also, he was the secretary's personal agent within the army, an informer who always reported everything to his master. Louvois even requested that inexperienced intendants such as Saint Pouenges check on the generals, for he wrote to that intendant in 1670: "You will see by the copy of my letter to Marshal Créquy what the king ordered me to tell him. Please take care to inform me if he executes punctually those intentions of His Majesty that I have made known to him."[7]

Since the intendants usually possessed copies of royal letters to the generals in addition to their own private instructions, they might in-

6. Rousset, *Histoire de Louvois*, 1:307, quotes this statement of Créquy.
7. Louvois to Saint Pouenges, 19 September 1670, A.G. A¹ 252, pp. 35r-37r.

deed warn Louvois if a general had overstepped his bounds or failed to obey orders. Consequently the intendants frequently incurred the rancor of the generals. In the incident between Créquy and Charuel, the general complained about the "contestations that have always arisen between Sieur Charuel and the generals with whom he has worked." To be fair, it must be said that the army intendant's main task was not spying, but rather that of being a troubleshooter, ready to report if anything went wrong and to intervene to correct the situation. The army intendant, armed with the secretary's confidence, was the ideal man for this task.[8]

Although the development of the army intendancy might seem retarded compared to that of the provincial intendancy, this development suited a basic need. The army intendant remained a simple agent with commission ready to execute any command from the secretary. His power base could rest only on the support of the minister. It was to the crown's advantage that army intendants did not develop the independence of the provincial intendant, with his fixed residence, permanent bureau, local tradition, and, eventually, ties with the native establishment. Army intendancies always remained more flexible, attracting new men as well as the powerful officeholding families. The war department drew from this pool of talent and molded instruments for its own purpose—service to the crown.

8. Créquy to the king, 25 October 1670, A.G. A¹ 250, pp. 157r–59v. Cf. Turenne's remark concerning intendant Charuel, quoted in Rousset, *Histoire de Louvois*, 1:492—"M. Charuel passe pour un homme dont les écritures sont fort dangereuses."

Appendix

Intendants of the Army

1630

Paul Hay du Châtelet and J. d'Estampes de Valençay   Hay was "intendant de justice" in the army in Italy (Savoy) cojointly with army intendant Valençay (Source: Hanotaux, *Origines des intendants*, pp. 122n1, 333).

Gaspard Du Gué   "intendant de nos finances, vivres et magazins" in the army of Italy commanded by the king in person and in his absence by Richelieu (Source: commission A.G. A[1] 12, fol. 131).

Charles Le Roy de La Potherie and Dreux d'Aubray   the "intendance de la justice et finances tant en son pais de Provence qu'en l'armée que Sa Majesté y envoie pour chastier la rebellion de ceux de la ville d'Aix" (Hanotaux, *Origines des intendants*, p. 295).

1632

René de Voyer d'Argenson   "intendant de la justice police et des finances pres de nre ch. Cousin [the prince of Condé] en nosd. pays de Limousin, de la Hauthe et basse Marche, dAuvergne et autres provinces larrovoisines et adjacentes" (comm[on] A.G. A[1] 14, fol. 32).

1633

François Sublet de Noyers   "intendant de la justice et police de nostred armee" in Picardy commanded by Marshal de Chaulnes (comm[on] A.G. A[1] 14, fol. 113).

Geoffrey Luillier d'Orgeval   "intendant de la justice et police en nostre provinces de Picardie Champagne et Isles de France." He had the care of the cavalry in garrison on the frontier (comm[on] A.G. A[1] 14, fol. 121).

Isaac Laffemas   "intendant de la justice police et finances tant en nosd. armées que dicelles de lad. province de Champagne, Metz, Toul, Verdun et autres lieux de nre obeissance et protection et par tout ailleurs ou nos armées se pourront estendre" (comm[on] A.G. A[1] 14, fol. 44).

Guillaume Bordeaux   "intendant de nos finances" in Marshal de La Force's army (comm^on A.G. A¹ 14, fol. 87).

1634

Guillaume Bordeaux   "intendant de noz finances vivres et magasins de nozd. armées d'Allemagne, estants ou qui seront soubs la conduite de nosd. Cousins les Mar.^aux de La Force et de Brezé" (comm^on A.G. A¹ 21, fols. 86–87).

Robert Arnauld d'Andilly   According to his memoirs, he was intendant of Brézé's army in Germany in 1634. His intendancy began near the end of 1634, so perhaps he was Bordeaux's replacement.

1635

Guillaume Bordeaux   "intendant de nos finances vivres et magasins en noz provinces de Champagne ["Bourgogne" had been crossed out] et autres voisines, et en noz armées et troupes qui y sejoureront" (comm^on A.G. A¹ 26, fol. 64).

Jacques Dyel de Miromesnil   "intendant de la justice police et finances de nostred. armée commandees par . . . les Mar.^aux Chastillon et Breze" (comm^on A¹ 26, fol. 43).

Jean de Choisy   'intendant de nre justice et police sur nos armées et troupes" in Champagne (comm^on A.G. A¹ 26, fol. 49).

René de Voyer d'Argenson   "intendant de la justice, police, finances et vivres en nred. armée qui nous commanderons en personne" (comm^on A.G. A¹ 26, fol. 52*bis*).

François Lasnier   "Le sieur Lanier, conseiller d'état et maître des requêtes, est fait intendant de justice et des finances dans l'armée du duc de Rohan" (27 October 1635 in *Table ou abrégé de la Gazette de France*, 2:312).

Charles Machaut   "intendant de nre justice police et finances en nred. province et armée de Provence" (comm^on A.G. A¹ 25*bis*, fol. 491).

1636

Louis Le Maistre de Bellejamme   "intendant de justice, police et finances en nostre dicte armee et province de Picardie" (Livet, *Intendance d'Alsace*, p. 53).

François Lasnier   "intendant de nre justice en nre armée et nre ambassadeur aux Grisons" serving with the duke of Rohan (A.G. A¹ 32, fols. 64–66).

Charles Machaut   "intendant de la justice police et finances en nostred armee de Bourgogne" commanded by the prince of Condé (comm^on A.G. A¹ 32, fol. 83).

Gaspard Du Gué   "intendant de nos finances en nostred. pays de Provence et en nos armées qui sont ou seront cy apres en nostred. province et en costes d'icelle" under Marshal de Vitry (comm^on A.G. A¹ 32, fol. 51).

Claude Gobelin   "intendant de la justice police et finances en nostred. armée" commanded by the duke of Orléans (comm^on A.G. A¹ 32, fol. 142).

René de Voyer d'Argenson   Notation in A.G. A¹ 32, fol. 178: "Le XII septembre 1636 il a esté fait deux commissions l'une pour M. Gobelin pour servir en l'armée de Monsieur et lautre po. M d'Argenson en celle de M. le Mar.^al de Force dIntend^t de la justice police et finances."

Olivier Le Fèvre d'Ormesson   "intendant de la justice police et de finances en notre prov^ce de Picardie et sur nos trouppes estant en lad. province" (A.G. A¹ 32, fol. 228).

Jacques Bigot   "intendant de la justice police et finances en mon armée de Champagne" (A.G. A¹ 88, fol. 8).

Louis Le Maistre de Bellejamme   "intendant de la justice police et finances en mon armée de Picardie" (A.G. A¹ 88, fol. 10).

Claude Gobelin   "intendant de nre justice police et finances en nre armée dAllemagne" (A.G. A¹ 88, fol. 5).

1637

Jacques Dyel de Miromesnil   "intendant de la justice, police et finances en l'armée commandée par nre cher cousin le duc de Longueville," dated 26 January when Dyel was intendant of the province of Normandy (comm^on A.G. A¹ 42, fol. 37). There is also a first draft of a commission for Dyel as intendant of the army in Burgundy commanded by Longueville and dated 26 March 1637 (A.G. A¹ 42, fol. 80).

Alexandre St Julien de Sève   "intendant de la justice police et finances en nostred. armee" commanded by Chastillon (comm^on A.G. A¹ 42, fol. 81). There is an accompanying notation that two commissions "semblables en blanc" were to be made out for the armies of marshals La Force and Brézé.

François Lasnier   "intendant de la justice police et finances en nre armée" commanded by La Meilleraye, grand master of the artillery (comm^on A.G. A¹ 42, fol. 128).

1638

François Lasnier   "intendant de la justice police et finances ["en nred. province" had been crossed out] en nos armees de Picardie" (comm^on A.G. A¹ 49, fol. 162).

René Voyer de Paulmy, sieur de Dorée   When Argenson, the regular army intendant in Italy, was recalled to court to consult with the king, Dorée was named to act in his absence as "intendant de la justice police et finances sur les troupes dans leurs garnisons, et a la campagne." Cardinal La Valette commanded the army (comm^on A.G. A¹ 49, fol. 196).

1639

Jacques Amelot de Beaulieu   "intendant de la justice police et finances en nred. armée de Guyenne" commanded by Condé and the archbishop of Bordeaux (comm^{on} A.G. A^1 56, fol. 46).

Claude Gobelin, Jean de Choisy, and Nicolas Bretel de Grémonville   Curious first draft for three commissions as "intendant de la justice police et finances en nostred. armee commandée par nostred. cousin [here follow three names: 1) Feuquières, 2) Chastillon, 3) the grand master of the artillery]." Gobelin was to serve in the army on the frontier of Picardy, Choisy on the frontier of Champagne, and Grémonville on the frontier of Champagne and Picardy (comm^{on} A.G. A^1 56, fol. 82).

Jacques Dyel de Miromesnil   "intendant de la justice police et finances et vivres en nostred. armee qui sera commandée par nostred. cousin le duc de Longueville" in Savoy-Italy (comm^{on} A.G. A^1 56, fol. 100).

Nicolas Bretel de Grémonville   A second commission for Grémonville as "intendant de la justice police et finances en nred. armée et province du Champagne, Brie, pays des trois evesches, Barrois et autres lieux ou l'estendre les logement d'icelle armée." This was for the winter quarters of La Valette's army (comm^{on} A.G. A^1 56, fol. 213).

1640

Claude Gobelin   "intendant de la justice, police, finances et vivres en nred. armée" commanded by La Meilleraye (comm^{on} A.G. A^1 62, fol. 121).

Nicolas Bretel de Grémonville   "intendant de la justice police finances et vivres en nre armée qui sera commandée par nosd. cousins les Mar.^{aux} de Chaunes et de Chastillon" (comm^{on} A.G. A^1 62, fol. 126).

Sieur Ollier   "intendant la justice police finances et vivres en nred. armee et province de Guyenne." Condé commanded the army (comm^{on} A.G. A^1 62, fol. 136).

François Cazet de Vauxtorte   "intendant de la police justice et finances en nred. pays de Provence et terres adjacentes, et en nre armee que nous faisons assembler en nred. province" (comm^{on} A.G. A^1 62, fol. 139).

René Voyer de Paulmy, sieur de Dorée   Dorée received another commission to have the direction of finances in the army of Italy during Argenson's absence (A.G. A^1 62, fol. 191).

Michel Le Tellier   Le Tellier replaced Argenson as "intendant de la justice police et finances et vivres en nred. armee d'Italie" (comm^{on} A.G. A^1 62, fol. 193).

Jacques Amelot de Beaulieu   "intendant de la justice police et finances en nosd. armee de Guyenne tant de terre que de mer" commanded by Condé and the archbishop of Bordeaux (comm^{on} A.G. A^1 62, fol. 305).

1641

René de Voyer d'Argenson   "intendant de la justice, police, finances et vivres . . . de nre armee et dans le pays de Catalongne" (comm^{on} A.G. A^1 67, fol. 76).

Claude Gobelin   "intendant de la justice police et finances en l'armee de Picardie" (comm^on A.G. A¹ 67, fol. 136).

Jean Lauzon   "intendant de la justice police et finances de nred. armees de Guyenne et Bearn" commanded by Schomberg (comm^on A.G. A¹ 67, fol. 101).

Jacques Amelot de Beaulieu   "intendant de la justice police et finances en nred. armee" commanded by the king in person (comm^on A.G. A¹ 67, fol. 157).

Jacques Dyel de Miromesnil   "intendant de la justice police finances et vivres en nred. armee d'Allemagne" commanded by the duke of Longueville. Miromesnil replaced Choisy as intendant (comm^on A.G. A¹ 67, fol. 158).

Sieur de la Court   "intendant des finances et vivres ["justice et police" had been crossed out] en lad. armee" of Lorraine (comm^on A.G. A¹ 67, fol. 180).

René Voyer de Paulmy, sieur de Dorée   "intendant des finances" in the army of Catalonia while Argenson remained at Barcelona. Brézé was the army commander (comm^on A.G. A¹ 67, fol. 234).

1642

Jean de Choisy   "intendant de la justice police et finances pres de nous et a nre suite et en larmee q. nous commandera en personne" (comm^on A.G. A¹ 71, fol. 29).

Claude Gobelin   "intendant de la justice police et finance en nred. armee de Roussilon" commanded by La Meilleraye (comm^on A.G. A¹ 71, fol. 49). The document noted that Gobelin had received a similar commission the previous year.

Sieur de la Court   "intendant des finances et vivres en lad. armée de Lorraine" (comm^on A.G. A¹ 71, fol. 51).

Louis Le Maistre de Bellejamme   "intendant de la justice, police et finances en nred. armée de Flandres" commanded by Harcourt (in the margin are the words "Champagne" and "Gremonville," implying that Grémonville received a similar commission) (comm^on A.G. A¹ 71, fol. 74). Bellejamme was provincial intendant in Picardy when he received this commission.

Jean de Choisy   Choisy, already intendant of the king's army, got an additional commission as "intendant de la justice police et finances en nre province de Champagne" for winter quarters (comm^on A.G. A¹ 71, fol. 147).

1643

Le Goux de la Berchère   "intendant de la justice police et finances en nos armees et province de Catalonge" as well as in Roussillon and Cerdergne (comm^on A.G. A¹ 78, 2d section of 3d part, fol. 1).

Michel Aligre de Saint-Lié   to "faire et exerciser la charge d'intendant des finances en lad armée" of Catalonia because La Berchère "se trouvant obligés de demeure en la ville de Barcelone" (comm^on A.G. A¹ 78, fols. 10–11). The army was commanded by La Mothe-Houdancourt.

Nicolas Bretel de Grémonville   "intendant de la justice, police, finances et vivres en nos armees d'Italie" to replace Le Tellier (comm^on A.G. A^1 79, 2d section, fol. 4).

Jean de Choisy   "intendant de la justice police finances et vivres en nostred. armée commandée par nos tres cher cousin le duc danguien [Enghien]" in Picardy (comm^on A.G. A^1 79, 2d section, fol. 6).

M. de Jeannin   "Intendant de la justice, police et finances et vivres en nos armees de Champagne" (comm^on A.G. A^1 79, 2d section, fol. 7). (In the margin is the notation of a commission for Champlastreux as intendant of the army in Burgundy under the command of La Meilleraye, but see the following item.)

Guillaume Bautre, count de Serran   The name of Champlastreux has been crossed out and Bautre's substituted as "intendant de la justice police finances et vivres en nre armée commandée par nred. Cousin le Mar.^al de La Meilleraye" on the frontier of Burgundy (comm^on A.G. A^1 79, 2d section, fol. 10).

Nicolas Fouquet and Jacques de Chaulnes   Joint commission as "intendants de justice police et finances en nostre province de Dauphine, et en l'armée troupes et garnisons qu'exerçoy en Icelle lesd. Srs de Seve et de Chazé" (comm^on A.G. A^1 80, fol. 82).

1644

François Villemontée   "intendant de la justice police et finances en nred. armee commandee par nred. tres cher oncle," the duke of Orléans, on the frontier of Picardy (comm^on A.G. A^1 86, fol. 127).

Jean-Edouard Molé de Champlastreux   "intendant de la justice police et finances en nre armee commandée par nostred cousin le duc danguien" (comm^on A.G. A^1 86, fol. 130).

Louis I Chauvelin   "intendant de la justice police et finances et vivres en nre armee d'Italie" to replace Grémonville (comm^on A.G. A^1 86, fol. 316).

1645

Pierre Goury   to "faire et exerciser la charge d'intendant des finances en lad. armée" of Catalonia commanded by Harcourt(comm^on A.G. A^1 92, fol. 236).

Pierre Imbert   "intendant de la justice, police et finances dans nre corps d'armée qui sera commandés par le Sr Comte Du Plessis Praslain." Imbert was intendant of Roussillon as well as the army of Catalonia (comm^on A.G. A^1 92, fol. 245).

François Cazet de Vauxtorte   "Intendant de la justice, police et finances dans l'Evesché de Spire, Archevesché de Mayence, Marquisat de Baden, Bas-Palatinat et autres pays, places tenues par nos armes en ces quartiers-là, pour en cette qualité avoir la direction et intendance de la justice police et finances à l'esgard de nos troupes et en toutes les choses concernant notre service à la conservation desdites villes et pays de notre obéissance" (comm^on printed in Livet, *Intendance d'Alsace*, pp. 925–27, and in Mousnier, *Lettres au Séguier*, 2:1071–75. Another copy of this document incorrectly dated "1633" is in A.G. A^1 14, fol. 110).

1646

Julien Pietre   to "faire et exercer lad. charge d'intend^t des finances en Icelle armée" during the illness of Chaulnes, intendant of the army of Flanders commanded by the duke of Orléans (comm^on A.G. A^1 96, fol. 148).

René de Voyer d'Argenson   "surintendant de la justice police finances et vivres en nostre armée de terre qui agira . . . avec nre armée navalle." Argenson's account of this expedition to Orbetello is printed in *Journal d'Ormesson*, 2:720–41 (comm^on A.G. A^1 96, fol. 138).

Jean-Edouard Molé de Champlastreux   "intendant de la justice police et finances en nosd. armée" commanded by the duke of Enghien (comm^on A.G. A^1 96, fol. 143).

Jacques de Chaulnes   "intendant de justice police et finances" in the army of Flanders commanded by the duke of Orléans (comm^on A.G. A^1 96, fol. 144).

Paul Le Gendre   "intendant des finances" in the army assembled on the frontier of Champagne and commanded by La Ferté Senneterre (comm^on A.G. A^1 96, fol. 147).

Jean de Choisy   "intendant de la justice police et finances en nostre armée de Champagne" commanded by the count of Soissons (comm^on A.G. A^1 96, fol. 156).

1647

Nicolas Fouquet   "intendant de la justice police et finances en nre armée commandée par nre tres cher ame et feal oncle le duc d'Orleans" assembled on the frontier of Picardy (comm^on A.G. A^1 101, fol. 153).

Jean-Edouard Molé de Champlastreux   "intendant de la justice, police et finances en notred. armée de Catalogne qui sera commandée par notred. cousin le prince de Condé" (comm^on A.G. A^1 103, pp. 124r–26r).

René Le Vayer   Royal letter to Le Vayer asking him to exercise the charge of intendant in the army of Flanders commanded by Marshals Gassion and Rantzau while awaiting the arrival of Fouquet as intendant (A.G. A^1 103, p. 354rv).

Paul Le Gendre   "intendant des finances en lad. armée" deployed on the frontier of Lorraine under La Ferté Senneterre (comm^on A.G. A^1 101, fol. 152).

Pierre Imbert   Royal letter to Imbert: "vous fassies toutes les fonctions de celle d'Intendant des finances comme si vous en avies le titre" in the army of Catalonia after Champlastreux's departure until the naming of a new intendant (A.G. A^1 104, pp. 96v–97v).

Jacques Brachet   "intendant des finances" at the Italian towns of Piombino and Porto Longone when a French expedition seized these places (A.G. A¹ 103, pp. 134r–38v).

Ennemond Servien   Various references in letters to Servien as intendant of the army of Italy under Prince Thomas of Savoy and "intendant de la justice et finances en Piedmont" (A.G. A¹ 104, pp. 284v–88r, and A¹ 106, fol. 199).

Pierre Goury   "intendant des finances et fortifications en notred. armee et places de Catalongne." He replaced Imbert, who returned to his intendancy of Roussillon (commᵒⁿ A.G. A¹ 104, pp. 314v–16v).

1648

Jean Balthazar   "intendant de la justice police et finances et vivres en notre armee de Lombardie" commanded by Du Plessis Praslain (commᵒⁿ A.G. A¹ 107, pp. 32r–35v).

Simon Arnauld d'Andilly (Pomponne)   "intendant de la justice police et finances en son armée de terre qui sera jointe a son armée navale" (Instructions, A.G. A¹ 107, pp. 240r–48v).

Ennemond Servien   Remained "intendant des mes armées" and ambassador to Savoy (A.G. A¹ 105, fol. 269).

Louis Gombault   Intendant of finance "en ces quartiers" when taken prisoner by the Spanish at Armentières (A.G. A¹ 110, fol. 232; A¹ 111, fol. 75). Was he an intendant of the army? Yes, according to B.N. *dossiers bleus* 319, he was "intendant des armées du Roy."

Claude Bazin de Bezons   "intendant de la justice police et finances" in the army of Catalonia, replacing Goury (commᵒⁿ A.G. A¹ 108, pp. 232r–34v).

1649

Jean de Choisy   "intendant de la justice police et finances en notred. armée et pres de la personne de notred. oncle" the duke of Orléans in Flanders (commᵒⁿ A.G. A¹ 114, pp. 26v–29r).

François Villemontée, Nicolas Fouquet, Le Goux de la Berchère, and Louis Laisné, sieur de la Marguerie   At the end of the above commission is this notation: "Il a este fait des sembables commissions pour les Sʳˢ de Villemontee Conᵉʳ d'Estat ordʳᵉ Fouquet et La Berchere Mᵉˢ des Requestes et laisné aussy Conᵉʳ d'Estat ordʳᵉ.

"Led Sʳ de Choisy ayant pour son departement le quartier du Roy ou commandoit Monseigneur le duc d'Orleans et sous luy Mgʳ le Prince et ceux de St Cloud, de Meudon, et de Surenne ou servi depuis led Sʳ de la Berchere dans lequel commandoit Monsʳ le Marᵃˡ de Grammont Lieutenant Genéral. Led. Sʳ de Villemontée au quartier de St Denis ou commandoit Monsʳ le Marᵃˡ Duplessis Praslain, led. Sʳ Fouquet a Lagny ou commandoit un marᵃˡ de camp, led Sʳ Laisné a Corbeil ou commandoit le Sʳ de Navailles Marᵃˡ de camp dans lequel quartier de Lagny et de Corbeil a commandé depuis le Sʳ Comte de Grancey en qualité de Lieutenant Genéral."

There may have been some shifting of these departments, for an order of the king to Fouquet began: "Sa Ma.^{te} ayant commis le Sr Fouquet con^{er} du Roy en ses conseils et M^e des Requestes ord^{re} de son hostel en la charge d'intendant de la justice police et finances en l'armée estant pres de sa personne pour resider specialement aux quartiers establis a Corbeil et autres lieux du pays de Brie aux environs et s'employer particulierement a faire des magasins pour la subsistance desd. troupes et a les faire vivre avec ordre" (A.G. A¹ 114, pp. 96v–98r).

Philibert Baussan   "intendant de justice, police, et finances en notred. armée lorsqu'elle sera en Champagne et ailleurs ou elle agira." These were Erlac's German troops called into France during the Fronde. Baussan remained intendant of Alsace and left the care of this province to Domiliers, a *commissaire des guerres,* during his absence (comm^{on} A.G. A¹ 114, pp. 211r–13v, and A¹ 114, pp. 199v–200v).

Jacques de Chaulnes   "intendant de la justice police et finances en nostre armée de Flandres" commanded by Harcourt (comm^{on} A.G. A¹ 113, fol. 201).

Philibert Baussan   Letter to Harcourt announcing that Baussan had been named "intendant de la justice police et finances et comm^{re} gnal" of the foreign troops serving in Luxembourg under Harcourt (A.G. A¹ 116, fol. 64).

Louis Le Tonnelier de Breteuil   "intendant de la justice police et finances a lesgard de nos troupes tant d'infanterie que de cavallerie françoises et estrangeres estant et qui seront cy apres en garnison dans nos places du Comté de Roussillon ou qui auront a y passer et sejourner." At the same time, Breteuil remained the intendant of Languedoc (comm^{on} A.G. A¹ 115, pp. 131r–34v).

1650

Nicolas Fouquet and Louis Laisné, sieur de la Marguerie   "intendant de la justice police et finances en notred. armée pres de notre personne" in Normandy. At the same time, "il a esté expedié aussy une pareille commission que celle cy dessus au Sr de la Marguerie . . . pour servir d'intendant en lad. armée de Normandie" (comm^{on} A.G. A¹ 120, pp. 104r–5v).

Jacques Paget   "intendant de la justice police et finances en l'armée de Champagne" commanded by La Ferté Senneterre (Instructions, A.G. A¹ 120, pp. 156r–59v).

Nicolas Fouquet   Royal letter to serve as intendant of the army of Burgundy commanded by the king. Fouquet did not need a new commission; he served "en qualité de M^e des Req^{tes} ord.^{res} de mon hôtel et en vertu de la prīnte, vous exerciez lad. Intendance de la justice police et finances en mad. armée avec les memes pouvoirs authoritez honneurs et facultez que vous pouriez faire si je vous en avois donnée ma commission."

Louis Laisné, sieur de la Marguerie   A notation to the above letter adds: "il a esté fait une semblable lettre que celle cy dessus a M. de la Margrie [*sic*] pour en qualité de Con^{er} d'estat servie aussy d'intendant en lad.

armee de Bourgogne cojointement aved led. Sr Fouquet" (A.G. A¹ 120, p. 177rv).

Antoine Bordeaux   "intendant de la justice en lad. armee" of Du Plessis Praslain in Flanders (A.G. A¹ 120, pp. 468rv, 504v–6r).

Charles Brèthe de Clermont   "intendant de la justice et finances en mon armée de Champagne" (A.G. A¹ 122, fol. 183).

Ennemond Servien   Various letters mention Servien as "mon ambassadeur en Piedmont et intendant de mon armee d'Italie" and "Intendant de la justice en mon armée d'Italie et dela les montes" (A.G. A¹ 122, fol. 429; A¹ 120, pp. 5r–6v).

1651

Simon Arnauld d'Andilly (Pomponne)   "intendant de la justice, police et finances en notred. armee de Catalongne" (comm^{on} A.G. A¹ 124, fol. 279).

Antoine Bordeaux   "intendant de la justice police et finances en nosd. armées" of Flanders, although the army assembled in Picardy and Champagne (comm^{on} A.G. A¹ 126, pp. 143r–45r).

Denis Marin   "intendant des finances en notred. armée que sera assemblée dans les occasions pñtes en notred. province de Guyenne et qui servira dans nos provinces ou il sera besoin de la faire agir" (comm^{on} A.G. A¹ 126, pp. 471r–73r).

Denis de Heere   "intendant de la justice police et finances en nostre armée qui sera assemblée pres de notre personne" against Condé (comm^{on} A.G. A¹ 125, fol. 194).

Claude Bazin de Bezons   "intendant de la justice police et finances en l'armée de Berry" (comm^{on} A.G. A¹ 125, fol. 196). A letter from Brienne to Vieuville in 1651 mentioned Bezons's service the preceding year, 1650, as intendant of the army (A.G. A¹ 127, pp. 86r–88r).

Antoine Bordeaux   "intendant de la justice police et finances en notred armée d'Italie" (comm^{on} A.G. A¹ 125, fol. 197).

Louis Boucherat   Letter to Boucherat "po. dire que l'on luy donne l'intendance de l'armée de Haute Guyenne" commanded by Saint Luc (A.G. A¹ 127, p. 234rv).

1652

Pierre de Pontac   Two variant commissions. The first, dated 12 April, named him "intendant de la justice, police et finances en notred. armee de Guyenne." The second, dated 18 May, named him "intendant de la justice et police ["finances" had been crossed out] en nostred. armée de Guyenne" (comm^{ons} A.G. A¹ 135, pp. 183r–84r, and A¹ 132, fol. 303).

Antoine Bordeaux   "intendant de la justice police et finances en nosd. armées servants pres de notre personne" under the command of Turenne and Hocquincourt near Compeigne and Corbeil (comm^{on} A.G. A¹ 135, pp. 210v–12r). In July of that year, a letter from the king named

Bordeaux intendant of the army commanded by La Ferté Senneterre, in addition to his being intendant of Turenne's army (A.G. A¹ 136, p. 6rv).

[Philippe?] Talon   "intendant des finances dans lad. corps d'armée" in Saintonge under the command of Du Plessis Bellièvre and Montauzier (comm^on A.G. A¹ 132, fol. 302).

Michel Aligre de Saint-Lié   Letter patent addressed to Aligre, "nre ame et feal Intendant des finances en nre armée de Catalongne. . . . mesmes en l'intendance des finances en l'armée de Catalonge que vous avez digne-ment exercee par le passe et laquelle nous vous avons donnée de nouveau" (A.G. A¹ 132, fol. 455).

Jean Ribeyre   "intendant des finances en nostred. armée de Guyenne et provinces voisines" commanded by the duc de Candalle. Ribeyre replaced Pontac (comm^on A.G. A¹ 132, fol. 304).

1653

Charles Brachet   "intendant des finances dans notred. armée de Lorraine" commanded by La Ferté Senneterre (comm^on A.G. A¹ 139, pp. 271r–73r).

Claude Bazin de Bezons   "intendant de la justice police et finances en notred. armée de Roussillon" commanded by Hocquincourt. Shortly afterward, Bezons became "visiteur général en Catalogne" (comm^on A.G. A¹ 138, fol. 400; A¹ 138, fol. 405).

[Philippe?] Talon   "intendant des finances de nred. armee" in Cham-pagne commanded by Turenne (comm^on A.G. A¹ 141, fol. 180).

Jean-Jacques de Mesmes, count d'Avaux   Two months later, Avaux was named "intendant de la justice police et finances de nred. armee com-mandee par nred. Cousin le mal. de Turenne." He also received a commis-sion for winter quarters of the troops in Soissons (comm^ons A.G. A¹ 141, fols. 181–82; A¹ 140, pp. 358r–59v). Hanotaux, *Origines des intendants,* p. 303n1, claimed that this had to be before 1642, when the said Mesmes died, but Edmond Esmonin, "Observations critiques sur le livre de M. Hanotaux: 'Origines de l'institution des intendants des provinces,'" re-jected Hanotaux's claim and reasserted the date 10 November 1653, in *Etudes sur la France des XVIIe et XVIIIe siècles,* p. 13.

Jean Colbert du Terron   "pour avoir l'intendance et direction du paye-ment de nosd. troupes qui seront en garnison dans nosd pays de Foix et Bigorre et frontieres de Guyenne" (comm^on A.G. A¹ 141, fol. 183).

Gédéon Tallemant   "intendant de la justice police et finances pres de nos troupes qui seront en quartier d'hiver et qui passeront et sejourneront en. . . . lad. generallité de Bordeaux" (comm^on A.G. A¹ 141, fol. 184).

Charles Machaut   A footnote to the above document noted that an iden-tical commission was sent to Machaut for the *generalité* of Montauban.

1654

Jean Colbert du Terron   Two variant commissions: that of 16 May named him "intendant des finances dans nre armée de Guyenne" commanded by the count d'Estrades; the second, dated 8 June, commissioned him "intendant des finances pres de nos troupes et dans le corps d'armee que destre composé et que sera com^dé du chef par le Sr Comte dEstrades" (comm^ons A.G. A¹ 142, fols. 236–37).

Charles Colbert de Croissy   "intendant de nos finances dans notred armée que sera embarquée en Provence po. faire descente es pays de ennemis" under the command of the duke of Guise and in his absence by Du Plessis Praslain (comm^on A.G. A¹ 143, pp. 362v–68r).

Pierre Gargan   Commission to have the direction of the subsistence and police of the troops in winter quarters in Amiens, Soissons, and Châlons (comm^on A.G. A¹ 144, pp. 207v–11v). The document noted his similar service the preceding year.

[Philippe or Jean?] Talon   "l'intendance et direction des deniers et danrées qui pourront este levées dans le pays ou les troupes de la gar^on du Quesnoy nouvellem^t reduittes sous lobeis^ce de Sa Ma^té les pourront estendre (comm^on A.G. A¹ 144, pp. 194v–99v).

Antoine Le Fèvre de la Barre   "intendant de la justice police et finances sur nos gens de guerre tant de cheval que de pied, françois et estrangers qui passeront logeront et sejourneront en notred. province de Dauphiné en corps ou autrement y feront leur levée assemblée et sy trouveront en quartier dhyver garnison ou en quelque sorte et maniere que ce soit" (comm^on A.G. A¹ 155, pp. 44r–45r). See also Esmonin, *Etudes sur la France*, p. 90.

François Bochart de Sarron Champigny   "intendant de la justice police et finances sur nos gens de guerre tant de cheval que de pied françois et estrangers qui hyvereront esd. pays de Bresse Bugey Vaulronney Gex et Dombs." He was already intendant of Lyonnais (comm^on A.G. A¹ 144, pp. 305r–9v).

1655

Jacques Brachet   "Intendant des finances" in the corps of the army of Italy commanded by the count of Broglie. He remained as intendant until his death in 1659 (comm^on A.G. A¹ 147, fol. 277).

Charles Colbert de Croissy   "pour en labsence dud. Sr de Bezons, avoir l'intendance de la police et finances en nostred armee de Catalongne" commanded by Conty. Bezons was intendant of both the province of Languedoc and the army of Catalonia and could not be in both places at once (comm^on A.G. A¹ 147, fol. 514).

1656

Louis Gombault   "intendant de la justice et police en notred. armée de Catalogne qui sera commandee en chef par . . . le Prince de Conty." The bishop of Orange, the visitor general, had charge of finances in the army (comm^on A.G. A¹ 147, fol. 516; Instructions, A.G. A¹ 149, pp. 151r–57v).

François Bochart de Sarron Champigny   "intendant de la justice police et finances sur nos gens de guerre tant de cheval que de pied françois et estranger qui sont et seront en quartier d'hyver dans lesd. diocezes et de Viviers Mende et le Puy." He was also intendant of Lyonnais (comm^on A.G. A¹ 147, fol. 285).

Antoine Le Févre de la Barre   "intendant de la justice police et finances sur nos gens de guerre tant de cheval que de pied françois estrangers qui seront en quartier dhyver de lad. generalité de Moulins" (comm^on A.G. A¹ 147, fol. 286).

François Villemontée   "intendant de la justice police et finances pres de nos troupes en la gñallitez de Soissons" (A.G. A¹ 150, pp. 248r–50r).

1658

Daniel Voisin   "intendant de la justice police et finances pres des troupes de Sa Ma^te en la province de Champagne" (A.G. A¹ 151, fol. 369).

François Bochart de Sarron Champigny   "Intendant de la justice, police et finances sur noz gens de guerre tant de cheval que de pied françois et estrangers qui passeront, logeront et sejourneront en nostredite province de Dauphiné, en corps ou autrement" (comm^on printed in Esmonin, *Etudes sur la France,* pp. 93–94).

1660

Louis Robert   A memoir concerning the expedition to Crete mentioned "M. Robert qui a este choisy pour avoir lintendance de cette armee" (A.G. A¹ 162, fol. 1).

[Philippe?] Talon   "intendant de l'armée de Flandres" (A.G. A¹ 164, pp. 44r–46r).

1663

Louis Robert   "intendant des nos finances et de la police dans les corps de trouppes que nous faisons passer dans les estatz de nred. cousin le duc de Parme" (comm^on A.G. A¹ 182, pp. 233–36).

   Amplification of this commission as "intendant de nos finances et de la police dans lad corps de troupes que nous faisons passer dans les estatz de nred. cousin le duc de Modene" (A.G. A¹ 182, pp. 442–44).

1664

Etienne Carlier   "intendant de la police et des finances sur le corps de troupes que nous avons envoyé soubs le commandement du Sr de Pradel" to Erfurt (B.N. *cabinet d'Hozier* 78, family "Carlier," fol. 2).

Honoré Courtin   "intendant de la justice police et finances en nre armée d'Italie" (comm^on A.G. A¹ 182, pp. 873–76).

Jacques Charuel   "intendant de nos finances, et de la police en notred.
armée de terre que sera jointe a notred. armée navalle" commanded by
the duke of Beaufort off the African coast (comm^on A.G. A¹ 184, fols.
329–30).

Louis Robert   "intendant de la police et finances" for the corps of troops
under Coligny sent to aid the Empire against the Turks in Hungary (Instructions, A.G. A¹ 189, pp. 108v–17r).

1665

Etienne Carlier   "intendant de nos finances et de la police dans ledit corps
de trouppes que nous faisons passer en Hollande" under the command of
Pradel (comm^on A.G. A¹ 198, no. 10, pp. 28r–29v).

1667

Etienne Carlier   "intendant du corps de troupes que Sa Ma.^te fait assembler du coste de la frontiere de Champagne"; Carlier served in Créquy's
army (A.G. A¹ 206, fol. 137).

Louis Robert   "intendant de l'armee de M. le Mar^al D'Aumont" (A.G.
A¹ 209, fol. 29).

Charles Colbert de Croissy   "intendant de l'armée" of Turenne (A.G. A¹
209, fol. 275).

1668

Louis-François Le Fèvre de Caumartin   Louvois announced: "vous avez
esté choisy par le Roy pour faire les fonctions d'Intendant de l'armée
d'Allemagne qui sera commandée la campagne prochaine par Monseigneur
le Prince" (A.G. A¹ 222, fol. 189).

Louis Robert   Robert received permission to go the army as intendant
(A.G. A¹ 222, fol. 293).

Paul Barillon d'Amoncourt   "intendant de la justice police et finances en
l'armée de Sa Ma^te" (A.G. A¹ 222, fol. 483).

1669

Sieur de la Croix   "intendant de la justice police et finances sur ledit corps
d'armée commandé par nostre dit Cousin le Duc de Navailles" on the
expedition to succor the Venetians against the Turks in Crete (comm^on
A.G. A¹ 238, pp. 33v–36v).

1670

Gilbert Colbert de Saint Pouenges   "intendant desdites troupes" sent to
Lorraine under the command of Créquy (A.G. preface to A¹ 252, unpaged).

Jacques Charuel   "intendant de police finances et vivres de l'armée de Sa Ma^te en Lorraine, et des villes et Chllmes de Courtray et Ath." He retained his intendancy in Flanders (A.G. A¹ 250, pp. 170r–71r).

1671

Louis Robert   "intendant de la justice, police et finances" of the troops sent under the command of Chamilly to assist the elector of Cologne at the end of the year (A.G. A¹ 272, pp. 423–25).

1672

Michel Le Peletier de Souzy   "pour faire les fonctions dIntendant dans le corps de troupes doit commander M. de Nancré." He remained intendant of Flanders (Instructions, A.G. A¹ 272, pp. 830–41).

Louis Robert   "pour faire les fonctions d'intendant dans l'armée de M. le Prince" (Instructions, A.G. A¹ 272, pp. 842–50).

Jacques Charuel   Letter of Louvois to Créquy: "Le Roy a choisy M. Charuel pour estre Intendant dans l'armée que vous commanderez pendant cette campagne" (A.G. A¹ 272, pp. 734–37).

Paul Barillon d'Amoncourt   "Intendant de justice police et finances dans l'armee de Sa Maiesté" (Turenne's army) (A.G. A¹ 293, fol. 46).

Louis Robert   With the invasion of Holland, Robert entitled himself "intendant de justice police et finances dans toutes les places et pays conquis par Sa Majesté en Holland." He remained with the army in Holland, however (A.G. A¹ 295, fol. 38).

1673

Barillon; Germain-Michel Camus de Beaulieu; Robert   In March 1673, Barillon returned home to handle personal affairs, and Charuel replaced him as "intendant de l'armée de mond. Cousin le Vicomte de Turenne" (A.G. A¹ 314, pp. 279v–80r). But Charuel and Turenne did not get along, so the crown replaced Charuel in July with Camus de Beaulieu (A.G. A¹ 316, p. 25v). Meanwhile, Robert remained with the army in Holland.

1674

Louis Robert; Michel Le Peletier de Souzy   Robert was "intendant de la justice police et finances en nred. armée commandée en chef par nostred. Cousin le Prince de Conde" (comm^on A.G. A¹ 367, fol. 7). However, Croquez, *Intendance de la Flandre wallonne*, p. 301, lists Le Peletier de Souzy as intendant in the army of M. le Prince and mentions that he served in this capacity at the battle of Senef.

Louis Machaut   The king named this intendant of Soissons: "intendant de la justice police et finances en nostredite armée commandée en chef par nostredit cousin le Vicomte de Turenne" in Germany (comm^on A.G. A^1 379, fol. 292).

Etienne Carlier   "intendant de la justice police et finances sur les troupes dont nostredite armée que nous faisons assembler sur nostre frontiere de Roussillon sera composée soit quelle soit en corps ou en quartiers" commanded by the count de Schomberg. At the same time, Carlier was intendant of the province of Roussillon (comm^on A.G. A^1 379, fol. 312).

1675

Germain-Michel Camus de Beaulieu   "intendant de la justice police et finances sur les troupes estants dans nostred. armée de Roussillon" commanded by Schomberg. The previous year Camus de Beaulieu had been intendant of contributions in the Franche Comté. He replaced Carlier as intendant of the army of Roussillon (comm^on A.G. A^1 432, pp. 442–44).

François Bazin   "intendant de la justice police et finances de nostred. armée commandée en chef par nostred. Cousin le Vicomte de Turenne" (comm^on A.G. A^1 432, pp. 647–53).

Louis Robert   Although there is no commission extant for Robert, he once again served as intendant of the army (Instructions for the supply of the army at Tournai, A.G. A^1 433, pp. 168r–70v).

Jean Colbert du Terron   "intendant de la justice police et finances sur nos troupes estant a Messine et autres places de Sicille" (A.G. A^1 636, fol. 231).

Auguste-Robert de Pomereu   "intendant de justice, police et finances en l'armée de Bretagne pour M. de Pomereu" (Canal, "Essai sur Auguste-Robert de Pomereu, intendant d'armée en Bretagne [1675–1676]," *Annales de Bretagne* 24 [1908–9]: 497–513).

1676

Germain-Michel Camus de Beaulieu   Because Carlier's age "ne luy permettoit plus de pouvoir agir avec la vigueur qu'il convient dans la conjoncture presente des affaires," the king replaced him with Camus de Beaulieu, who combined "l'intendance de la province et de l'armée de Roussillon" commanded by the duke of Navailles (A.G. A^1 482, pp. 73–75).

François Bazin   "intendant en l'armée d'Allemagne" commanded by the duke of Luxembourg (A.G. A^1 483, pp. 500–501).

Louis Robert; Michel Le Peletier de Souzy   Robert remained intendant of the army of Flanders; Croquez, however, in *Intendance de la Flandre wallonne*, p. 301, names Le Peletier de Souzy as intendant of Créquy's army in Flanders. Both served in the army, but it is difficult to determine the precise title of each, for their commissions are lacking.

1677

Louis Robert   "Sa Ma^te ayant choisy led. Sr Robert pour servir d'intendant en l'armée qu'Elle a resolu d'assembler pour faire le siege de St Omer" (A.G. A¹ 531, pp. 335–46). Other letters refer to him as "intendant de l'armée de Monsieur."

Charles Camus du Clos   "pour faire les fonctions d'intendant des corps que Sa Ma^te a resolu de faire agir en Milanois pendant la Campagne prochaine" (A.G. A¹ 533, pp. 688–99).

François Bazin   Bazin remained the intendant of the army of Germany commanded by Créquy, then by Schomberg.

1678

Charles Mouceau de Nollant   "intendant de l'armée commandée par M. de Schomberg" (A.G. A¹ 582, p. 198r).

Louis Robert   "Mr Robert intendant de l'armee du Roy" commanded by the duke of Luxembourg (A.G. A¹ 582, p. 174v, in margin).

François Bazin   Intendant of the army of Germany under Créquy (A.G. A¹ 609, fols. 20–21).

1683

Michel Le Peletier de Souzy   "intendant de la justice, police et finances et des armées de Sa Ma^té en Flandre" (A.G. A¹ 701, no. 22).

Louis II Chauvelin   Chauvelin, intendant of the Franche Comté, was named "intendant dans le camp . . . assembler sur le Saône" commanded by the marquis de Boufflers (A.G. A¹ 708, p. 22).

1684

François Le Tonnelier de Breteuil   "Ce meme jour Mons. de Louvois a escrit un mot a Mons de Breteuil intendant en Flandre, pour luy donner advis que le Roy la nommé pour estre Intendant de l'armée de Sa Ma^té" in Flanders under the command of Boufflers (A.G. A¹ 723, fol. 34).

1687

Michel-Louis Malézieu   "intendant des camps et armees du Roy" (B.N. *carrés d'Hozier* 404, family "Malézieu," fol. 272; also mentioned in A.G. A¹ 791, unpaged).

1688

Thomas Heisse   Named by the king as intendant of the troops sent to Cologne (A.G. A¹ 818 passim; see table at front of volume).

1689

Dreux-Louis Du Gué de Bagnols   "de Bagnols intendant de l'armée" commanded by Marshal d'Humières (A.G. A¹ 864, fol. 206).

Claude de La Fond   "intendant de l'armee" of Germany (commᵒⁿ A.G. A¹ 872, no. 161).

Thomas Heisse   "intendant des troupes du Roy dans le pais de Cologne" commanded by the marquis de Sourdis (A.G. A¹ 886, preface, unpaged).

Jean-Etienne Bouchu   Intendant of the army of Italy in addition to his provincial intendancy of Dauphiné (see Esmonin, *Etudes sur la France*, p. 100).

1690

Claude de La Fond   "intendant de la justice police et finances en nostre dite armée commandée en chef par nostred. fils le dauphin, et soubs luy par nostred. Cousin le Marᵃˡ de Lorge." La Fond was also intendant of the Franche Comté (commᵒⁿ A.G. A¹ 936, unpaged).

Dreux-Louis Du Gué de Bagnols   "intendant de l'armée de Flandre" commanded by Marshal de Luxembourg (A.G. A¹ 941, unpaged; see letters of Louvois, 7 January and 11 April 1690).

Jean-Etienne Bouchu   "intendant de la justice police et finances en nostredite armée" of Piedmont commanded by Catinat. Bouchu was also intendant of Dauphiné (commᵒⁿ A.G. A¹ 1006, fol. 59).

Raymond Trobat   "intendant en Roussillon, et de l'armée de Sa Majesté en Catalogne" commanded by the duc de Noailles (A.G. A¹ 1103, introduction).

René Joüenne d'Esgrigny   "intendant des troupes du Roy en Ireland" under the command of M. de Maumont (A.G. A¹ 960, unpaged, letter of Louvois, 16 February 1690).

1691

Dreux-Louis Du Gué de Bagnols   Letter of Louvois: "je vous adresse la commission necessaire pour faire la fonction d'intendant dans l'armée du Roy en Flandres" commanded by the duke of Luxembourg (A.G. A¹ 1044, no. 8).

Jacques-Louis Le Marié   "intendant de l'armée du Roy" near the Moselle commanded by Boufflers (A.G. A¹ 1044, no. 18).

Claude de La Fond   Letter of Louvois announcing the sending of a commission "qui vous est necʳᵉ pour faire la fonction d'intendant dans l'armée du Roy en Allemagne" (A.G. A¹ 1085, unpaged, letter of Louvois, 2 May 1691).

Raymond Trobat   "intendant de l'armée du Roy en Roussillon, pour

Mons. le President Trobat du mois d'avril 1691" (A.G. A^1 1103, unpaged, letter of Louvois, 2 May 1691).

Jean-Etienne Bouchu "intendant de l'armée du Roy en Italie pour M. Bouchu pendant la campagne de 1691" (comm^{on} A.G. A^1 1077, no. 183).

Death of Louvois   16 July 1691

Bibliography

MANUSCRIPT SOURCES

Agen: Archives Départementales de Lot-et-Garonne.
Series EE. 21, Ville d'Agen.

Paris: Archives des Affairs Etrangères (A.A.E.).
Mémoires et documents FRANCE: 796, 813, 815.

Bibliothèque Nationale (B.N.).
Cabinet des Titres:
1) *cabinet d'Hozier:* 78 (Carlier family), 196 (Imbert), 223 (Malézieu), 260 (Paget), 316 (Talon).
2) *carrés d'Hozier:* 180 (Chauvelin), 283 (Gargan), 348 (Imbert), 404 (Malézieu), 455 (Mouceau, Du).
3) *collection de Chérin:* 192 (Talon).
4) *dossiers bleus:* 16 (Amelot), 58 (Barillon), 66 (Baussan and Bautru), 69 (Bazin), 96 (Bigot), 112 (Bordeaux), 115 (Boucherat), 116 (Bouchu), 128 (Brachet), 133 (Bretel), 150 (Camus), 154 (Carlier), 171 (Charuel), 178 (Chaulnes), 187 (Choisy), 203–4 (Colbert), 219 (Courtin), 267 (Fèvre, Le), 273 (Fonds, La), 305 (Gargan), 309 (Gendre, Le), 319 (Gombault), 326 (Goury), 327 (Goux, Le), 336 (Gué, Du), 353 (Héere), 370 (Joüenne), 375 (Laffemas), 413 (Machaut), 428 (Marin), 445 (Mesmes), 476 (Mouceau, Du), 523 (Pietre), 535 (Pontac), 539 (Potherie), 564 (Ribeire), 569 (Robert, Robert à Paris, and Robert, De la Fortelle), 613 (Servien), 623 (Tallemant), 627 (Le Tellier), 660 (Vayer, Le), 677 (Voisin).
5) *nouveau d'Hozier:* 286 (Robert).
6) *pièces originales:* 598 (Carlier and Carlier en Picardie), 694 (Charuel), 995 (Desgrigny), 1063 (Esgrigny), 1378 (Goury), 1502 (Heiss), 1853 (Marié), 2499–2500 (Robert), 2885 (Trobat).
Cinq-cents Colbert: 499.
Fonds français: 3,758, 4,014, 14,018.
Recueil Cangé: F168, boite K, tome X.

Vincennes: Archives de la Guerre (A.G.).
Series A¹: 12, 13, 14, 21, 25*bis*, 26, 32, 42, 49, 56, 62, 65, 67, 71,

78, 79, 80, 86, 88, 91, 92, 96, 101, 103, 104, 105, 106, 107, 108, 110,
111, 113, 114, 115, 116, 118, 120, 121, 122, 124, 125, 126, 127, 132,
133, 134, 135, 136, 137, 137*bis*, 138, 139, 140, 141, 142, 143, 144,
145, 147, 149, 150, 152, 155, 162, 163, 164, 168, 169, 170, 172, 178,
181, 182, 184, 189, 190, 198, 206, 208, 209, 210, 214, 222, 223, 224,
225, 226, 227, 238, 250, 252, 272, 292, 293, 294, 295, 296, 314, 316,
338, 339, 350, 367, 379, 432, 433, 482, 483, 531, 533, 582, 609, 636,
701, 708, 723, 791, 864, 872, 886, 936, 941, 960, 1006, 1044, 1077,
1085, 1103.

PRINTED SOURCES

Arnauld d'Andilly, Robert. *Mémoires de Messire Robert Arnauld d'Andilly.*
Collection des mémoires relatifs à l'histoire de France, edited by Petitot,
2d series, vols. 33–34. Paris, 1824.
Choiseul du Plessis Praslain, César. *Mémoires des divers emplois et des
principales actions du maréchal du Plessis.* Nouvelle collection des
mémoires relatifs à l'histoire de France, edited by Michaud and Poujoulat,
vol. 31. Paris, 1857.
Colbert, Jean-Baptiste. *Lettres, instructions et mémoires de Colbert.* Edited
by Pierre Clément. 8 vols. Paris, 1861–82.
Coligny-Saligny, Jean, comte de. *Mémoires du comte de Coligny-Saligny.*
Société de l'histoire de France, edited by Monmerqué. Paris, 1841.
Cosnac, Gabriel-Jules, comte de, ed. *Souvenirs du règne de Louis XIV.*
8 vols. Paris, 1866–82.
Dangeau, Philippe de Courcillon, marquis de. *Journal du marquis de
Dangeau.* Edited by Soulié, Dussieux, Mantz, Montaiglon, de Chennevières,
and Feuillet de Conches. 19 vols. Paris, 1854–60.
*Estat et gouvernement de France: Comme Il est depuis la majorité du roy
Louis XIV, à present régnant.* 7th ed. of 1648 version. Amsterdam, 1653.
Feuquières, Antoine de Pas, marquis de. *Mémoires de M. le marquis de
Feuquiere, lieutenant général des armées du roi; contenans ses maximes
sur la guerre, et l'application des exemples aux maximes.* New ed., rev. and
cor. London, 1736.
Guignard, Pierre-Claude de. *L'Ecole de Mars ou Mémoires instructifs sur
toutes les parties qui composent le corps militaire en France, avec leurs
origines, et les differentes maneuvres ausquelles elles sont employées.*
2 vols. Paris, 1725.
Hermite-Soliers, Jean-Baptiste de l', and Blanchard, François. *Les Eloges de
tous les premiers présidens du Parlement de Paris, depuis qu'il a esté rendu
sédentaire jusques à présent. Ensemble leurs généalogies, épitaphes, arms
et blazons, en taille douce.* Paris, 1645.
*Histoire du temps ou le véritable récit de ce qui s'est passé dans le Parlement
depuis le mois d'aoust 1647 jusques au mois de novembre 1648. Avec les
harangues et les advis différents, qui ont esté proposez dans les affaires
qu'on y a solennellement traittées.* [Paris], 1649.
Horric de Beaucaire, Charles-Prosper-Maurice, comte de, ed. *Savoie-
Sardaigne et Mantoue.* Recueil des instructions données aux ambassadeurs
et ministres de France depuis les traités de Westphalie jusqu'à la révolution française, vol. 14, pt. 1. Paris, 1899.
Journal contenant tout ce qui s'est faict et passé en la cour de Parlement de

Paris, toutes les chambres assemblées, sur le suiet des affaires du temps present. Paris, 1652.

La Bruyère, Jean de. *Oeuvres de La Bruyère.* Edited by Gustave Servois. 3 vols. Paris, 1865–78.

Le Fèvre d'Ormesson, Olivier. *Journal d'Olivier Lefèvre d'Ormesson et extraits des mémoires d'André Lefèvre d'Ormesson.* Collection de documents inédits sur l'histoire de France, edited by Pierre-Adolphe Chéruel. 2 vols. Paris, 1860–61.

Le Peletier, Claude. *Deux Mémoires historiques de Claude Le Pelletier.* Edited by Louis André. Paris, 1906.

Louis XIV. *Mémoires de Louis XIV pour l'instruction du dauphin.* Edited by Charles Dreyss. 2 vols. Paris, 1860.

Mazarin, Jules, cardinal de. *Lettres du cardinal Mazarin pendant son ministère.* Collection de documents inédits sur l'histoire de France, edited by Pierre-Adolphe Chéruel and Georges d'Avenel. 9 vols. Paris, 1872–1906.

Mousnier, Roland, ed. *Lettres et mémoires addressés au chancelier Séguier (1633–1649).* Publication de la Faculté des lettres et sciences humaines de Paris, series "Textes et documents," tome 6. 2 vols. Paris, 1964.

Nodot, François. *Le Munitionaire des armées de France, qui enseigne à fournir les vivres aux troupes avec toute l'oeconomie possible.* Paris, Brussels, and Lyons, [1697].

Richelieu, Armand-Jean du Plessis, cardinal de. *Lettres, instructions diplomatiques et papiers d'état du cardinal de Richelieu.* Collection de documents inédits sur l'histoire de France, edited by Denis-Louis-Martial Avenel. 8 vols. Paris, 1853–77.

―――. *Testament politique.* Edited by Louis André. Paris, 1947.

Tallemant des Réaux, Gédéon. *Les Historiettes de Tallemant des Réaux.* Edited by Monmerqué and Paris. 3d ed. 9 vols. Paris, 1854–60.

Talon, Omer. *Mémoires.* Nouvelle collection des mémoires relatifs à l'histoire de France, edited by Michaud and Poujoulat, vol. 30. Paris, 1857.

Tavannes, Jacques de Saulx, comte de. *Mémoires de Jacques de Saulx comte de Tavannes suivis de l'histoire de la guerre de Guyenne par Balthazar.* Edited by Célestin Moreau. Paris, 1853.

SECONDARY SOURCES

Aboucaya, Claude. *Les Intendants de la marine sous l'ancien régime.* Paris, 1958.

André, Louis. *Michel Le Tellier et l'organisation de l'armée monarchique.* Paris, 1906.

―――. *Michel Le Tellier et Louvois.* Paris, 1942.

Aumale, Henri d'Orléans, duc d'. *Histoire des princes de Condé pendant les XVIᵉ et XVIIᵉ siècles.* 8 vols. Paris, 1863–96.

Avenel, Georges, vicomte d'. *Richelieu et la monarchie absolue.* 2d ed. 4 vols. Paris, 1895.

Barbier, Alfred. *Notice biographique sur René de Voyer d'Argenson, intendant d'armée du Poitou, ambassadeur à Venise (1596–1651).* Poitiers, 1885.

Bernard, Leon. *The Emerging City: Paris in the Age of Louis XIV*. Durham, N.C., 1970.

Bluche, François. *Les Magistrats du Parlement de Paris au XVIIIe siècle (1715–1771)*. Annales littéraires de l'Université de Besançon, vol. 35. Paris, 1960.

———. *L'Origine des magistrats du Parlement de Paris au XVIIIe siècle*. Mémoires de la Fédération des sociétés historiques et archéologiques de Paris et de l'Ile-de-France, tomes 5–6. Paris, 1956.

Canal, Séverin. *Les Origines de l'intendance de Bretagne: Essai sur les relations de la Bretagne avec le pouvoir central*. Paris, 1911.

Caron, Narcisse-Léonard. *Michel Le Tellier: Son Administration comme intendant d'armée en Piémont, 1640–1643*. Paris and Nantes, 1880.

Charmeil, Jean-Paul. *Les Trésoriers de France à l'époque de la Fronde: Contribution à l'histoire de l'administration financière sous l'ancien régime*. Paris, 1964.

Chéruel, Pierre-Adolphe. *Histoire de France pendant la minorité de Louis XIV*. 4 vols. Paris, 1879–80.

———. *Mémoires sur la vie publique et privée de Fouquet surintendant des finances d'après ses lettres et des pièces inédites conservées à la bibliothèque impériale*. 2 vols. Paris, 1862.

Corvisier, André. *L'Armée française de la fin du XVIIe siècle au ministère de Choiseul: Le Soldat*. 2 vols. Paris, 1964.

Couyba, L. *Etudes sur la Fronde en Agenais*. 3 vols. Villeneuve-sur-Lot, 1901–3.

Croquez, Albert. *L'Intendance de la Flandre wallonne sous Louis XIV (1667–1708)*. Lille, 1912.

Doolin, Paul Rice. *The Fronde*. Cambridge, Mass., 1935.

Doucet, Roger. *Les Institutions de la France au XVIe siècle*. 2 vols. Paris, 1948.

Dublanchy, Charles-Nicolas. *Une Intendance d'armée au XVIIIe siècle: Etude sur les services administratifs à l'armée de Soubise pendant la guerre de sept ans*. Paris, [1905].

Esmonin, Edmond, ed. *Etudes sur la France des XVIIe et XVIIIe siècles*. Publication de la Faculté des lettres et sciences humaines, Université de Grenoble. Paris, 1964.

Frondeville, Henri de. *Les Conseillers du Parlement de Normandie au seizième siècle (1499–1594)*. Société de l'histoire de Normandie. Paris and Rouen, 1960.

———. *Les Présidents du Parlement de Normandie (1499–1790)*. Société de l'histoire de Normandie. Paris and Rouen, 1953.

Girard d'Albissin, Nelly. *Genèse de la frontière franco-belge: Les Variations des limites septentrionales de la France de 1659 à 1789*. Bibliothèque de la Société d'histoire du droit des pays flamands, picards et wallons, no. 26. Paris, 1970.

Godard, Charles. *Les Pouvoirs des intendants sous Louis XIV particulièrement dans les pays d'élections de 1661 à 1715*. Paris, 1901.

Goubert, Pierre. *Cent Mille Provinciaux au XVIIe siècle: Beauvais et le Beauvaisis de 1600 à 1730*. Paris, 1968.

———. *Louis XIV et vingt millions de Français*. Paris, 1966.

Gruder, Vivian R. *The Royal Provincial Intendants: A Governing Elite in Eighteenth-Century France*. Ithaca, N.Y., 1968.

Hanotaux, Gabriel. *Origines de l'institution des intendants des provinces d'après les documents inédits.* Paris, 1884.

Henry, l'Abbé. *François Bosquet, intendant de Guyenne et de Languedoc, évêque de Lodève et de Montpellier: Etude sur une administration civile et ecclésiastique au XVIIᵉ siècle.* Paris, 1889.

Kossmann, Ernst H. *La Fronde.* Leiden, 1954.

Lair, Jules-Auguste. *Nicolas Fouquet, procureur général, surintendant des finances, ministre d'état de Louis XIV.* 2 vols. Paris, 1890.

Legrand-Girarde, Emile-Edmond. *L'Arrière aux armées sous Louis XIII: Crusy de Marcillac, évêque de Mende, 1635–1638.* Paris, 1927.

Livet, Georges. *L'Intendance d'Alsace sous Louis XIV, 1648–1715.* Publication de la Faculté des lettres de l'Université de Strasbourg, fascicule 128. Paris, 1956.

Lottin, Alain. *Vie et mentalité d'un Lillois sous Louis XIV.* Lille, 1968.

Mandrou, Robert. *La France aux XVIIᵉ et XVIIIᵉ siècles.* Nouvelle Clio, no. 33. 2d ed., rev. Paris, 1970.

Marion, Marcel. *Dictionnaire des institutions de la France aux XVIIᵉ et XVIIIᵉ siècles.* 1923. Reprint. Paris, 1969.

Mentz, Georg. *Johann Philipp von Schönborn Kurfürst von Mainz Bischof von Würzburg und Worms, 1605–1673: Ein Beitrag zur Geschichte des siebzehnten Jahrhunderts.* 2 vols. Jena, 1896–99.

Mercier, Ernest. *Histoire de l'Afrique septentrionale (Berbérie) depuis les temps les plus reculés jusqu'à la conquête française (1830).* 3 vols. Paris, 1888–91.

Mongrédien, Georges. *Le Bourreau du cardinal de Richelieu, Isaac de Laffemas.* Paris, 1929.

Moote, A. Lloyd. *The Revolt of the Judges: The Parlement of Paris and the Fronde, 1643–1652.* Princeton, N.J., 1971.

Mousnier, Roland. *Etat et société en France aux XVIIᵉ et XVIIIᵉ siècles:* [Part 1] *Le Gouvernement et les corps.* Les Cours de Sorbonne. Paris, n.d.

———. *Peasant Uprisings in Seventeenth-Century France, Russia, and China.* Translated by Brian Pearce. New York and Evanston, 1970.

———. *La Vénalité des offices sous Henri IV et Louis XIII.* 2d ed., rev. and enl. Paris, 1971.

———, ed. *Le Conseil du roi de Louis XII à la Révolution.* Publication de la Faculté des lettres et sciences humaines de Paris-Sorbonne, series "Recherches," tome 56. Paris, 1970.

Navereau, André-Eugène. *Le Logement et les ustensiles des gens de guerre de 1439 à 1789.* Poitiers, 1924.

Parent, Michael, and Verroust, Jacques. *Vauban.* Paris, 1971.

Parker, Geoffrey. *The Army of Flanders and the Spanish Road, 1567–1659.* Cambridge Studies in Early Modern History, no. 3. Cambridge, 1972.

Porchnev, Boris. *Les Soulèvements populaires en France de 1623 à 1648.* Paris, 1963.

Prevost and Roman d'Amat. *Dictionnaire de biographie française.* 12 vols. to date. Paris, 1933–.

Ranum, Orest A. *Richelieu and the Councillors of Louis XIII: A Study of the Secretaries of State and Superintendants of Finance in the Ministry of Richelieu, 1635–1642.* Oxford, 1963.

Roberts, Michael. *Essays in Swedish History.* Minneapolis, Minn., 1967.

Rousset, Camille. *Histoire de Louvois et de son administration politique et militaire.* 4 vols. Paris, 1862–63.

Roux, Marie, marquis de. *Louis XIV et les provinces conquises.* Paris, 1938.

Sanabre, José. *La Acción de Francia en Cataluña en la pugna por la hegemonía de Europa (1640–1659).* Barcelona, 1956.

Saulnier, Frédéric. *Un Prélat au XVIIᵉ siècle: François de Villemontée, évêque de Saint-Malô (1660–1670), sa femme et ses enfants.* Extrait des mémoires de la Société archéologique d'Ille-et-Vilaine. Rennes, 1903.

Shennan, J. H. *The Parlement of Paris.* Ithaca, N.Y., 1968.

Sicard, François. *Histoire des institutions militaires des Français.* 4 vols. Paris, 1834.

Table ou abrégé des cent trente-cinq volumes de la Gazette de France, depuis son commencement en 1631 jusqu'à la fin de l'année 1765. 3 vols. Paris, 1766–67.

Zeller, Gaston. *L'Organisation défensive des frontières du nord et de l'est au XVIIᵉ siècle.* Paris, 1928.

———. *Les Temps modernes: De Louis XIV à 1789.* Histoire des relations internationales, vol. 3, under the direction of Pierre Renouvin. Paris, 1955.

Articles

André, Louis. "Le Maréchal de la Mothe-Houdancourt (son procès, sa rébellion, sa fin)." *Revue d'histoire moderne* 12 (1937): 5–35, 95–125.

Bluche, François. "L'Origine sociale des sécretaires d'état de Louis XIV (1661–1715)." *XVIIᵉ siècle,* nos. 42–43 (1959), pp. 8–22.

Bordes, Maurice. "Les Intendants de province au XVIIᵉ et XVIIIᵉ siècles." *L'Information historique,* May–June 1968, pp. 107–21.

Buisseret, David. "Les Précurseurs des intendants de Languedoc." *Annales du Midi,* n.s., no. 86 (1968), pp. 80–88.

———. "A Stage in the Development of the French *Intendants:* The Reign of Henri IV." *Historical Journal* 9 (1966): 27–38.

Canal, Séverin. "Essai sur Auguste-Robert de Pomereu, intendant d'armée en Bretagne (1675–1676)." *Annales de Bretagne* 24 (1908–9): 497–513.

Cilleuls, Jean des. "Un Grand Intendant d'armée: Moreau de Séchelles (1690–1760)." *Revue de l'intendance militaire,* no. 26 (1953), pp. 39–79.

———. "Le Service de l'intendance à l'armée de Rochambeau." *Revue historique de l'armée,* numéro spécial (1957), pp. 43–61.

Corvisier, André. "Les Généraux de Louis XIV et leur origine sociale." *XVIIᵉ siècle,* nos. 42–43 (1959), pp. 23–53.

Cubells, Mme. "Le Parlement de Paris pendant la Fronde." *XVIIᵉ siècle,* no. 35 (1957), pp. 171–99.

Dent, Julian. "An Aspect of the Crisis of the Seventeenth Century: The Collapse of the Financial Administration of the French Monarchy (1653–61)." *Economic History Review,* 2d ser., 20 (1967): 241–56.

Loirette, Francis. "Un Intendant de Guyenne avant la Fronde: Jean de Lauson (1641–1648)." *Bulletin,* Comité des travaux historiques et scientifiques, section de philologie et d'histoire, 1957, pp. 433–61.

Milot, Jean. "Du Commissaire des guerres à l'intendant militaire." *Revue historique de l'armée,* numéro spécial (1968), pp. 39–48.

———. "Evolution du corps des intendants militaires (des origines à 1882)." *Revue du Nord* 50 (1968): 381–410.

Moulias, D. "Les Origines du corps de l'intendance." *Revue historique de l'armée*, 1967, pp. 83–88.

Mousnier, Roland. "Etat et commissaire: Recherches sur la création des intendants des provinces (1634–1648)." *Forschungen zu Staat und Verfassung: Festgabe für Fritz Hartung*, edited by Richard Dietrich and Gerhard Oestreich, pp. 325–44. Berlin, 1958.

———. "Note sur les rapports entre les gouverneurs de provinces et les intendants dans la première moitié du XVIIe siècle." *Revue historique* 228 (1962): 339–50.

———. "La Participation des gouvernés à l'activité des gouvernants dans la France du XVIIe et XVIIIe siècles." *Schweizer Beitrage zu Allgemeinen Geschichte* 20 (1962–63): 200–229.

———. "Quelques Aspects de la fonction publique dans la société française du XVIIe siècle." *XVIIe siècle*, nos. 42–43 (1959), pp. 3–7.

———."Quelques Raisons de la Fronde: Les Causes des journées révolutionnaires parisiennes de 1648." *XVIIe siècle*, nos. 2–3 (1939), pp. 33–78.

———. "Recherches sur les Syndicats d'*Officers* pendant la Fronde: Trésoriers généraux de France et élus dans la Révolution." *XVIIe siècle*, nos. 42–43 (1959), pp. 76–117.

Schmidt, Charles. "Le Rôle et les attributions d'un 'intendant des finances' aux armées, Sublet de Noyers, de 1632 à 1636." *Revue d'histoire moderne et contemporaine* 2 (1900–1901): 156–75.

Stiot, Robert-Desiré. "Le Commissariat des guerres: Son organisation—son évolution—ses attributions." *Revue administrative*, no. 59 (1957), pp. 454–63.

Trout, Andrew. "The Proclamation of the Treaty of Nijmegen." *French Historical Studies* 5 (1968): 477–81.

Index